POOR DESIGN

Poor Design:
An Invalid Argument Against Intelligent Design

Copyright © 2019 Jerry Bergman

Published in the United States by BP Books in Tulsa, Oklahoma.

Library of Congress Control Number: 2019933956

ISBN: 978-1-944918-16-3

For author inquiries please send email to info@bplearning.net.

Bookstore bulk order discounts are available. Please contact info@bplearning.net for more information.

1st printing

Poor Design

An Invalid Argument Against Intelligent Design

By Jerry Bergman

Endorsements

As I biologists teaching at a research University, I often come across these claims of poor design and have made a mental note to respond. The research pressures at my university always got in the way. Fortunately, Professor Bergman did what I have wanted to do for some time, and he did an admirable job and saved me a lot of work. This is a critical area because these poor design claims are misleading and can be harmful because they can result in inappropriate treatment in the medical area.

Students need to understand these irresponsible claims driven by ideology are wrong and, at the least, misleading. For this reason, it is my hope that this book becomes part of the curriculum of every medical school and every program that teaches anatomy and physiology to health science students. In short, this book effectively and accurately demolished the many claims of poor design in the humans body.

—**Willard Lake, Ph.D. (Retired)**

An important topic in the biological origins controversy, especially the evolution of humans, is the claims of poor design made by evolutionists that they attempt to argue disproves creation by an intelligent designer and proves evolution. Dr. Bergman documents with nearly 1,000 footnotes the fact that, in contrast to the claims of some, most all of these poor design claims are without foundation. A carefully evaluation of these claims reveals that actually the organs and structures discussed are, in fact, very well-designed and the suggested changes which the critics clam are better design are either poorer design or lethal designs. Professor Bergman in his discussions relied heavily on the peer reviewed medical literature written by well-respected researchers and published in mainline journals. This book covers the most well-known claims, including the backwards retina, the prostrate, the human back, the human knee and the human birth canal.

The issues presented in this book also effectively documents the fact that the Darwinian worldview can, and often does, cause scientists to naively accept conclusions based on very flimsily evidence. It also shows that evolution has not helped but, in the cases covered by Bergman, hindered science and, in no small number of cases, has also injured patients. This is the only book ever published that I know of that refutes the common poor design claim made by evolutionists. This much-needed book fills a large void in the literature. It is, unlike many others published in the field, a work that objectively evaluates the negative effect of the Darwinian worldview on science. The fact that those who do not know history are condemned to repeat the mistakes of history is well-illustrated in this important new work by a noted biology college professor and prolific author.

—Norman Geisler, Ph.D.

Dr. Norman Geisler is a long-time professor and author of some 80 books, many of which are on evolution, including *Origin Science, Knowing the Truth about Creation*, and the definitive book on the history of the creation-evolution debate titled, *Creation & The Courts: Eighty Years of Conflict in the Classroom and the Courtroom* (Crossway, 2007).

A few weeks ago I came across a short newspaper article in the Pittsburgh Post-Gazette on the claims of poor design of several body structures including, of course, the human spine, the knee, and the eye. Naturally, the explanation for the poor design was, as usual, evolution. Then I discovered Dr. Jerry Bergman has already taken up the challenge of answering this particular false narrative as to why we humans have back and knee aliments. Dr. Bergmans book, *Poor Design: An Invalid Argument Against Intelligent Design*, takes on each of the favorite examples of poor design and documents the fact that sound medical science demonstrates the fact that our back, knees and other supposedly poorly designed systems are actually extremely well-designed. For the doctrinaire evolutionists, everything that happens—good, bad, or in between—is attributed to evolution by natural selection of mutations. I am very pleased that Dr. Bergman has taken on the preposterous claims of poor design of our bodies and has carefully and provided a thorough, scientific approach to understanding our human ailments. If natural selection has failed to select optimal designs, then how did all the intricate features of the many different animals manage to get anything right? Since you will hear about "poor design" in the context of evolution, Dr. Bergmans latest book carefully and thoroughly refutes this spurious interpretation of body ailments and the indirect attack on our Creator as revealed in the Bible.

—Theodore J. Siek, Ph.D. (biochemistry)

Jerry Bergman is again at his best in providing useful reference materials to support the case that the diversity of life on earth is masterfully designed. In this resource, Dr. Bergman takes on the mantra of evolutionists that the human body is "poorly designed." He shows that critics like Jerry Coyne, Abby Hafer, Richard Dawkins, Kenneth Miller, and Francis Ayala all demonstrate a profound lack of knowledge of the structures they fault. As we have seen over the years, their criticisms are not backed up by people who actually conduct research on those body parts. Dr. Bergman digs into the primary literature to show that every critique about something in the human body being outright dysfunctional by these celebrity evolutionists and many other wannabes is scientifically incorrect. In doing so, Dr. Bergman shows that their mistake exceeds mere expressions of ignorance. By their arrogant assertions that the bodys design isnt what a sensible human engineer would do, these evolutionists audaciously intend to mock God. Their smug ridicule of human anatomy and their claims that it is exhibit A of poor design is now embarrassingly exposed as a conceited scientific blunder.

—**Randy J. Guliuzza, M.P.H. (Harvard),**
M.D. (University of Minnesota)

I have considered myself a "theistic evolutionist," but that may change following my reading of Dr. Bergmans latest and possibly his most important book: *Poor Design: An Invalid Argument Against Intelligent Design.* Ive read several of Professor Bergmans books in the past. He had *almost* convinced me that his position is correct, but now he has done it in this massive, extremely well-researched tome that nearly overtasked my computer printer. Psalms 19:1, 2 reads: "The heavens declare the glory of God; and the firmament sheweth his handiwork. Day unto day uttereth speech, and night unto night sheweth knowledge." This scripture says to look around you for proof that God has created all you see. As a Medical Doctor, I can appreciate both this scripture and the anatomical insight his book imparts. I have worked with Dr. Bergman in the past on several chapters in this book, so I can affirm the effort he put in this book to ensure medical and anatomical accuracy.

—**Thomas B. Stogdill, M.D.**

Contents

Acknowledgments

I wish to thank Kevin McLeod, M.D.; Clifford Lillo, M.S., John Woodmorappe, M.A., Clifford Lillo, M.A., and especially Bert Thompson, Ph.D., Melanie Steinke, P.T., Sara Gilbert, P.T., Scott Gilbert, P.T., David Demick, M.D., Vij Sodera M.D., William Morris M.D., Jody Allen R.N. and E. Norbert Smith Ph.D., for their helpful comments and their insight on an earlier draft of this manuscript

For help on the chapter on the inverted retina (Chapter 10), I thank Dr. Tara Richmond, O.D., Clifford Lillo, M.S., John UpChurch; David Stoltzmann, Optical Engineer; Don DeYoung, Ph.D. physicist, Lindley-Anderson for producing the illustrations, and George F. Howe for editorial assistance and Joseph Calkins, M.D., currently an ophthalmologist in private practice, and formally Assistant Professor of Ophthalmology at Johns Hopkins University in Baltimore, Maryland. His M.D. is from the University of Michigan Medical School in Ann Arbor, MI. and Jody Allen, R.N. for their comments on an earlier draft of this chapter.

For help on the vas deferens chapter (Chapter 8) foremost is E. van Niekerk who worked very closely with me on this article, but also Professor Tony Jelsma, Ph.D. and Mark C. Biedebach Ph.D., Professor Emeritus of Physiology, Department of Biological Sciences, California State University, Long Beach, Dr. Loren Wanner, M.D. Obstetrician-Gynecologist, and, last but not least, Professor Holbird Doyle, PhD. Last, I thank professor Raymond Sleiman who earned his M.D. at St. Joseph University and has been Board certified in urology since 1981. He did his residency at Good

Samaritan Hospital and St. Louis University Hospital. Dr. Sleiman is a Clinical Professor at Keck School of Medicine of the University of Southern California. He also has been in private practice for 38 years in Long Beach, California and at Tri-City Regional Medical Center, Los Alamitos Medical Center, and Long Beach Memorial Medical Center and has 25 hospital affiliations.

For the help on the Left Recurrent Laryngeal Nerve (Chapter 9), I thank Vij Sodera, M.D., J. Y. Jones, M.D., and Clifford Lillo, M.A., and MaryAnn Stewart, M.A.

The background illustration used in chapter titles was licensed from Shutterstock.com and was drawn by Jorgen Mcleman.

Chapter 1

Introduction

This book is a response to the common argument by Darwinists that the human body is very poorly designed, thus, in the critics mind, proving it was not designed but rather has evolved. Actually, of the 10 organ systems and the over 200 organs, depending on what structures are defined as organs, only at most a handful are even claimed to be poorly designed. Nonetheless, the following claim is typical. This claim, flush with mocking Intelligent Design, says creation did not

> happen the way Charles Darwin supposed it did. It's designed. Intelligently. Dunno by who. Maybe God, who knows? Somebody intelligent, though. As though anybody who is a human being and thus has a human being's spinal column, digestive system, and genitalia... who would run a sewer pipe through a playground .... could possibly believe in good faith that we were designed by an intelligent creator.[1]

In doing research on examples of this claim for this book, I found, with few exceptions, almost always close to the same 10 or 12 were listed as poorly designed. Most of the poor design claims are answered by the research reviewed in this volume. As an example, the website titled *Top*

---

[1]Brayton, Tom. 2016. Teach the Controversy: A review of *Expelled; No Intelligence Allowed*. https://www.alternateending.com/2016/06/teach-the-controversy.html.

10 Design Flaws in the Human Body: From our knees to our eyeballs, our bodies are full of hack solutions[2] is fairly representative of criticisms of the human body, and the majority of the criticisms are covered in this volume.

An example is at the 2013 Annual AAAS meeting, consisting of the most prestigious scientists in the world. After acknowledging we humans "are the most successful primates on the planet," they added that our bodies would not win many awards for good design. This claim was "the consensus of a panel of anthropologists who described in often-painful (and sometimes personal) detail just how poor a job evolution has done sculpting the human form." One member added:

> The scientists showed how the very adaptations that have made humans so successful—such as upright walking and our big, complex brains—have been the result of constant remodeling of an ancient ape body plan that was originally used for life in the trees. "This anatomy isn't what you'd design from scratch," said anthropologist Jeremy DeSilva of Boston University. "Evolution works with duct tape and paper clips.[3]

This book responds to this claim, effectively documenting the opposite—our bodies, even those parts that evolutionists claim are poorly designed, are in fact *marvelously* well designed.

A Note on the Chapters

Each chapter was written to stand alone, thus some material is repeated, such as the material in the chapter on the retina (Chapter 10) and the chapter on the blind spot (Chapter 11).

---

[2]Chip Rowe, May 12, 2015.
http://nautil.us/issue/24/error/top-10-design-flaws-in-the-human-body.
[3]Ann Gibbons. 2013. Human Evolution: Gain Came with Pain. *Science*. Feb. 16, 2013.
http://www.sciencemag.org/news/2013/02/human-evolution-gain-came-pain.

Part I

Design of the Musculoskeletal System

Chapter 2

The Spinal Column

Darwinism has misled researchers into developing a set of therapies that have proved detrimental to treat certain back problems. These therapies were based on the Darwinian conclusion that humans evolved from primates that walked on all fours, and that back problems were primarily the result of complications resulting from humans' evolved upright posture.

In short, Darwinism teaches that our vertebral structure originally evolved to walk on all fours, and back problems exist today largely because humans now walk upright on vertebrae that originally evolved to walk quadrupedally. This theory has led to a treatment protocol that is now recognized as impeding healing, and has caused enormous pain and suffering. The orthodox treatment used today is, in many ways, the opposite of the older, now disproven, Darwinism-influenced theory.

A new field called Darwinian medicine is attempting to redefine medicine according to evolutionary perspectives. Nesse and Williams claim that this new view will enable medical practitioners to cure, or at least to better understand, disease.[1] The current enthusiasm by many to apply Darwinian conclusions to medicine is contraindicated by past attempts to use Darwinism to determine medical practice. One of the best examples that illustrates how these conclusions have adversely affected medicine is the application of Darwinism to the treatment of back pain.

---

[1]Nesse and Williams, 1994.

This is an important example because up to 90% of all Americans suffer at least one debilitating episode of back pain during their lives.[2]

Darwinists have taught for decades that the reason for this problem was bipedalism was superimposed by evolution on a skeleton previously well-adapted for quadrupedal motion.[3] In the words of Krogman: "Although man stands on two legs, his skeleton was originally designed for four. The result is some ingenious adaptions, not all of them successful."[4] The late Harvard evolutionist Ernst Mayr concluded that the "human species is highly successful even though it has not yet completed the transition from quadrupedal to bipedal life in all its structures."[5] What changes he expects as the evolutionary transition is completed he does not say. Krogman concludes that when humans started walking upright "a terrific mechanical imbalance" resulted, and backaches consequently became common.[6]

Krogman also opined when our pre-human ancestors walked on all fours, the skeleton was arched like a cantilever bridge with the trunk and abdomen representing the load suspended from the half-circle weight-balanced arch. The main bridge was a jointed, crane-like extension (the neck), and the force balance was achieved throughout the system *only* when walking on all fours. Tattersall is even more dogmatic, concluding that our past "is always with us, as the suffering of those with slipped discs—the inevitable result of adapting a quadrupedal spine to upright posture—attests."[7]

Krogman claims that the advantages of this cantilever system were lost when humans started walking upright and the backbone was forced to accommodate itself "to the new vertical weight-bearing stresses." He concludes that evolution accomplished this in humans by breaking up the single curved back arch into the s-curve design we now possess. Krogman adds that humans are born with the simple ancestral curved arch, but at

[2]Goldmann and Horowitz, 2000.

[3]Gracovetsky, 1996; Krames Communications, 1986.

[4]Krogman, 1951, p. 54.

[5]Mayr, 2001, p. 282.

[6]Krogman, 1951.

[7]Tattersall, 1998, p. 203. The word disk is often spelled disc when describing the human back.

the age of four months, when we begin to hold our heads erect, a new anterior sloping curve called lordosis (the opposite direction is a kyphosis curve) develops in the neck region (the cervical spine).

The Evolution Theory of a New Lordosis Curve in the Lumbar Spine

The evolution theory concluded that we also evolved a new lordosis curve in the lumbar spine, but the upper trunk and pelvic region (the thoracic) still retain a kyphotic lumbar curve.[8] Many evolutionists view this lumbar spine curvature, often called a swayback, "as an imperfect adaptation in man's supposed struggle to progress from four-footed stance to two-footed stance."[9] In the words of one popular author, back trouble began when our "ancestors decided to stand erect. Instead of a nicely balanced suspension bridge" the back became "a tent pole."[10] This view was explained in some detail by Vertosick as follows:

> How did this unfortunate anatomy evolve? ...Why do we have the disease-prone spines that we do? ...the basic layout of the spine evolved a hundred million years ago for animals maintaining a largely horizontal posture. ...evolving humans simply took a horizontally designed backbone and jacked it upright with little modification.

The backbone of a Segosaur is basically the same as ours. An example he provides is

> the stegosaurus (or the cow for that matter), the enormous weight of the animal was directed *perpendicular* to the axis of the spine, the body weight hanging down from the backbone like wet laundry from a clothesline, putting little pressure on the discs. In the human design, our body weight is directed

---

[8]Tuttle, 1972.
[9]Smail, 1990, p. 20.
[10]Ratcliff, 1975, pp. 227-228.

Figure 2.1: Basic Vertebrae Bone Segments

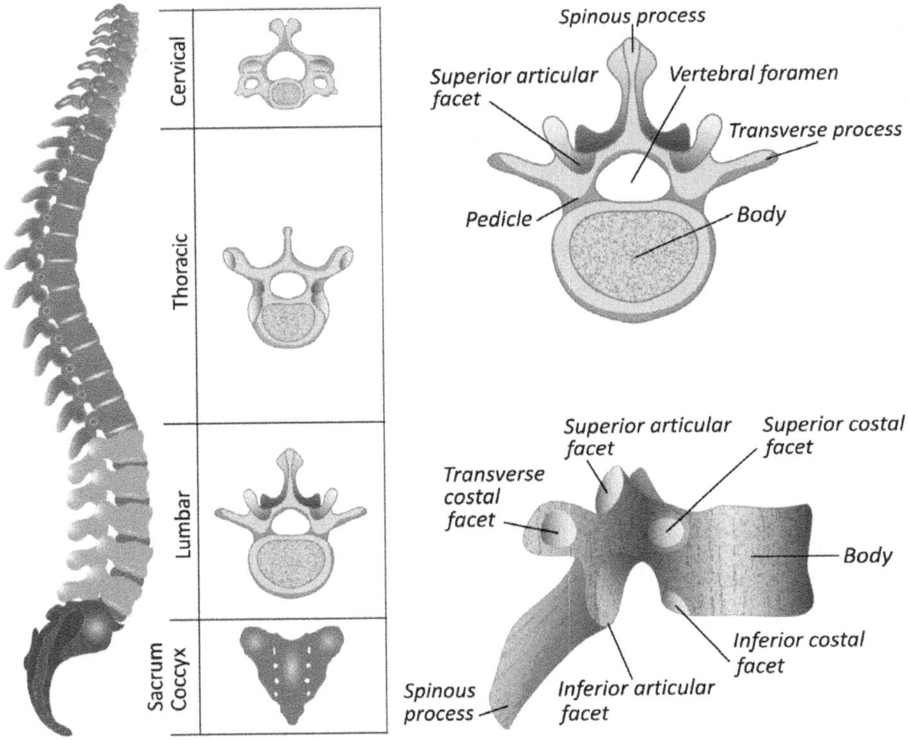

Image Credit: Ellen Bronstayn / Shutterstock.com

parallel to our spinal axis, a radical change. Using a horizontal spinal architecture for vertical walking is like using a screwdriver to drive nails: the nails may get driven, but a lot of screwdrivers will be shattered in the process.

He then concludes that the "price we pay for our new posture" is high because humans have

> only been using our newly upright spines for a few hundred thousand years—a mere blink of a Darwinian eye—there has been no time for natural selection to work out all of the kinks inherent in such a radical reorientation of the spine. Thus, I see sciatica as the curse of early hominids like us. Fortunately, evolution didn't abandon us entirely. Our erect posture damages our spines but also frees our hands, giving us one way to compensate for our rupturing discs: we can now become surgeons and remove them.[11]

Vertosick concludes his theory, which assumes modern human evolution from an ape ancestry, that our back design was a trade-off because

> evolution balanced the benefits of unencumbered forelimbs against the sporadic incapacity produced by failing spines. The fact that our planet currently crawls with countless human bodies proves that this trade-off was worthwhile. Given another hundred million years, nature might yet work out the kinks in our spines, but, until then, we're stuck with jerry-rigged technology that has yet to keep pace with the demands placed on it.[12]

The problem with this future prediction that evolution "might yet work out" is that

> evolution has no mercy and doesn't care if we have pain so long as that pain doesn't permanently cripple us or kill us

---

[11] Vertosick, 2000, pp. 93-94.
[12] Vertosick, 2000, p. 110.

before we've had the opportunity to reproduce. Ruptured discs are painful but not lethal, and they tend to run their course in a few weeks. Moreover, bad discs are more of a problem for those of us beyond our peak childbearing years.[13]

Since Darwinists conclude our spine was "deformed" as a result of humans standing and walking erect, the logical treatment for back pain would be to decrease or, ideally, reverse the lordotic curve.[14] To accomplish this, physician Dr. Williams devised a series of exercises now called "Williams flexion exercises" to reduce lordosis that evidently was used widely in many medical back treatment programs for years (and is still used in some places) in spite of its limited success, partly because it was completely logical—from the evolutionists' paradigm. [15] However, Mooney claims there never has been a scientific study that demonstrated the effectiveness of this, or any other treatment based on the Darwinist theory.[16]

Despite widespread use of the flexion (bending forward) exercises to reduce lordosis, back pain remained a severe problem.[17] This approach often failed, and consequently, surgery was often resorted to in order to correct the problem. Surgery improved the condition *in less than half* of all cases, and many patients were worse off than before.[18] Improved techniques since then have improved this record somewhat, partly because of the use of techniques that are better able to identify candidates for surgery. However, the surgery approach still faces a significant number of less-than-ideal outcomes.

A New Theory Not Based on Darwinism

A Wellington, New Zealand, physical therapist named Robin McKenzie discovered posture exercises that *restored* full natural lordosis and actu-

---

[13] Vertosick, 2000, p. 110.
[14] Krogman, 1951.
[15] Williams, 1965; Ishmael and Shorbe, 1960.
[16] Mooney, 1985.
[17] Smail, 1990, p. 20.
[18] Berger, 2000.

ally *decreased* or even eventually abolished back pain in many patients.[19] This result was the exact *opposite* of what had been recommended by Dr. Williams and other therapists based on Darwinian explanations. McKenzie is not a creationist, "but his work supports the creationist view that the lumbar lordosis is not a deformity causing inherent strain due to past evolutionary development. The lumbar spine is, instead, the most efficient means for supporting weight and providing for movement in erect, bipedal posture."[20]

It is now widely recognized that back problems generally are not due to maladaptation caused by upright posture, but rather, to abuses of the body that are common in modern life. This includes lack of exercise, poor posture, stress, and the requirement that one be in awkward positions for long periods of time, such as bending forward on an assembly line or spending hours on a computer. In short, anything that *decreases* the natural lordosis and back health causes problems, the opposite of the Darwinian based approach.

Back problems are due largely to weakness caused by lack of exercise, a highly sedentary lifestyle, and in some cases inheritance. Other major factors that lead to back problems are osteoporosis, back bone deterioration, obesity, and smoking, which is a vasoconstrictor and also contributes to osteoporosis.[21] One indication that modern society is largely to blame for back problems is the finding by physicians that patients living in Third World countries rarely report chronic back pain.[22]

Modern Back Problem Treatments

The Darwin-based Williams theory also recommended extra bed rest to deal with back pain, the opposite of the solution taught to patients today. Exercise (especially brisk walking) and only normally required sleep, are now the mainstay of back therapy. Research by Sobel and Klein on the

---

[19] 1996; Jacob and McKenzie, 1986; McKenzie and Kubey, 1985.
[20] Smail, 1990, p. 21.
[21] Clements, 1968.
[22] Mooney, 1985.

Figure 2.2: Main Back Abnormalities

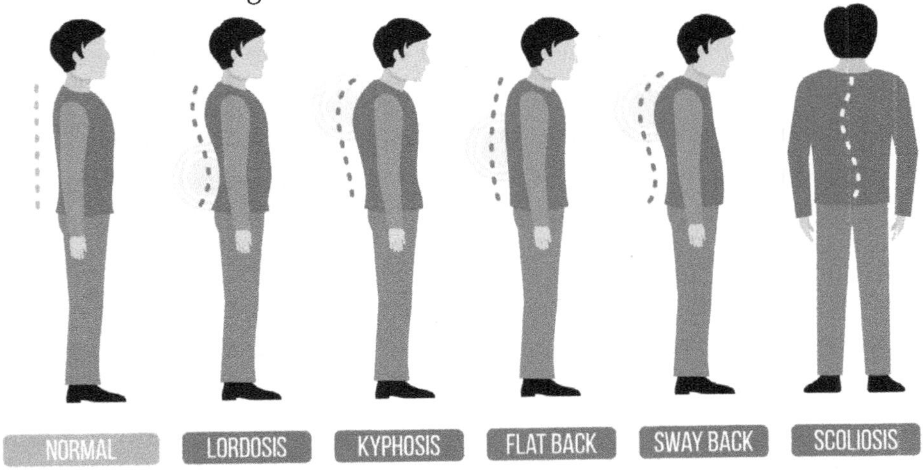

Image Credit: MSSA / Shutterstock.com

effectiveness of various back therapies has found that walking was highly beneficial in the vast majority of cases and in the long run helpful for 98 percent of the 500 cases they studied.[23]

Modern treatment is designed to improve both sitting and standing postures, educate patients in lifting mechanics to help them lift correctly, develop good sleep habits, use a firm-but-comfortable mattress and, most important, stay in shape with regular physical exercise, including both moderate muscle building and a stretching program.

Exercise is now an important part of both prevention and treatment not just for chronic, but also for acute, back pain.[24] Exercise improves muscle tone and allows the full set of muscles to carry the weight, and stretching improves flexibility, allowing the back system to function properly. Called the *biomechanical model*, it stresses that the body functions as an irreducibly complex unit, and must function as a harmonious set to work properly.[25] Cornell University medical school Professor Willibald Nagler concluded, based on the evaluation of close to 500 low-back-pain patients, that the

[23]Sobel and Klein, 2000, p. 136.
[24]Deyo, 1998, p. 52.
[25]Richardson, et al., 2002; Hides, et al., 1986.

vast majority of cases, even in cases of severe back pain can be treated by exercise.[26]

Especially important are the lumbar multifidus and the transverse abdominis muscles.[27] Controlled studies consistently show that exercise to strengthen these muscles significantly reduces back pain. In one long-term study, the recurrence rate was only 30 percent in the exercise group vs. 84 percent in the control group.[28] The Baptist Medical System Back School notes that research "has shown conclusively that exercise and correct body mechanics is the best prevention for back pain."[29] Slumping in front of a computer or television weakens muscles that support and protect the spine.

The conclusion that back problems are not due to our alleged evolutionary past is now accepted by evolutionists and creationists alike. Professor of osteopathic medicine, David Shuman, along with his coworker concluded that there is "no question ... the human back, given proper care and rightly understood, is an astonishingly effective mechanism. As much as the more frequently lauded human brain, the human back is the hallmark of our true nobility and a major factor in the ... supremacy of ... man."[30]

They conclude, "given proper care, . . . and just a little understanding, your back will take on any job you ask of it .... When it fails, in practically all of the more severe cases the failure is due to some sort of [muscle] weakness. " This "truly marvelous hunk of machinery, an amazingly durable arrangement ready to serve the purposes of a ditch digger or a banker, a prizefighter or a stenographer, equally well" requires only regular maintenance.[31]

The use of treatment therapies that re-establish normal lordosis have gradually found acceptance in the medical community.[32] Controlled scientific research has supported the McKenzie approach in comparison

---

[26]Sobel and Klein, 2000, pp. vii-viii.
[27]Young, 2004, p. 10.
[28]Hides, et al., 2001.
[29]1998, p. 3.
[30]Shuman and Staab, 1960, p. 12.
[31]Shuman and Staab, 1960, pp.13, 28.
[32]Murtagh, 1997.

with other approaches, including the old evolutionary-based Williams approach.[33] It is now widely recognized that, by teaching patients to maintain lumbar lordosis via back support and exercise, pain can be markedly reduced. In many cases, total healing can occur, even of a herniated disc.[34] To maintain a lordotic curve, many authorities now recommend moderate lordotic support by use of a long round pillow called a lumbar roll.[35] Many chairs and automobile power seats now have built-in lumbar support systems to achieve this goal.

In short, we used to "blame evolutionary design for our back problems"–specifically the "claim that the spinal column originally evolved to support people who didn't stand upright and walk on two legs."[36] We now know that "back problems result from our modern, sedentary style of living. This is good news–it means that you can prevent most non-congenital back problems by learning habits and taking actions that will help your back stay healthy."[37] In the words of Sobel and Klein: "Instead of dismissing back pain as 'the price we pay for walking upright,' specialists are at last coming to see back pain as the consequence of *not walking upright enough*—of spending too much time sitting at the desk, in front of the television, or behind the wheel."[38]

This insight is reflected, for example, in the training of Olympic weight lifters who learn to bend their backs while lifting in harmony with the McKenzie theory and in contradiction to the Darwinist-based Williams theory.[39] No doubt the therapy that developed from evolutionary assumptions has caused a huge amount of, not only back pain, but also permanent back damage, and likely has been one factor that has motivated surgery that may do more harm than good.[40] The Baptist Medical System Back School manual notes that one of "the *major* contributors to

---

[33]Razmjou, et al., 2000; Delaney and Hubka, 1999; Foster, et al., 1999; Samanta and Beardsley, 1999; Gillan, et al., 1998; Stankovic and Hohnell, 1995.

[34]Nelson, et al., 1999; Deyo, 1998.

[35]McKenzie, 1988.

[36]Yost, 1986, p. 4.

[37]Yost, 1986, p. 4.

[38]Sobel and Klein, 2000, p. 5

[39]Lamb, 2001, p. 1.

[40]Lamb, 2001, p. 8.

back pain" is stress and fatigue, and that one of the most common causes of lower back pain is "a sudden bending or lifting during a period of extreme tension or fatigue. The back is the focus of stress for some people. As soon as they are feeling pressure and anxiety about some event in their lives, the back responds by 'going out.'"[41]

Further research has also found that some muscles function to cause body movement, others serve to protect the integrity of the joint structure system. Research by Richardson, et al., found that a major problem often involved in back problems is a breakdown in the small intrinsic muscles of the spine and/or in the manner in which their activity is controlled.[42] Their study has resulted in the development of a highly effective exercise program to strengthen these muscles to treat back pain.

An evaluation of the cause of most common back problems also disproves the claim that back problems are a result of evolution from walking on all fours to an upright posture. Most people in American society spend a great deal of time sitting, and research has found that a *major cause* of back problems is *improper* sitting.[43] When standing erect, the lumbar portion of the back is lordotic. This inward curve is often lost to a large degree while sitting, producing a kyphotic or outward arch. This kyphosis produces increased pressure on the disc, which may be damaged by excessive or long-term sitting; especially problematic is unsupported sitting not using a backrest.

Other Research

Other research has found that forward tilting opens the angle between the hip and the upper torso, producing both a more relaxed posture and a more natural lordotic back posture. This is why the major recommendation to reduce back pain is walking, i.e. while walking a natural posture is usually assumed. In a weightless environment, or often when sleeping on one's side, a natural lordosis is achieved, reducing pressure on the discs.

---

[41]The Baptist Medical System Back School manual, 1998, p. 38, emphasis theirs.
[42]Richardson, et al., 1999, p. 4.
[43]Sanders and McCormick, 1993, p. 436.

This research also supports the conclusions that common back problems today are often due to our modern lifestyle. Techniques used to reduce disc pressure include back support, an arm rest and lumbar support.[44]

Another problem is excess postural fixity, i.e. sitting in one position for an extended period of time without major postural movement. The "human body is simply not made to sit in one position for long periods of time."[45] Reasons why include the fact that healthy discs depend on pressure changes to receive nutrients and remove waste products because fluids are exchanged only by osmotic pressure, not circulation as is true of most of the rest of the body. The prolonged effect of reduced nutritional exchanges contributes to disc degeneration.[46]

This problem can be reduced by walking around every few minutes, or even by using chairs that can be adjusted to a variety of positions, or those that allow the user to rock, to relieve postural fixity problems. Flexing and bending backward can also be very helpful as can movement adjustment to facilitate comfort, as well as periodic adjustment to achieve variety. For this reason, adjustable chairs are becoming more widespread.

Why the Williams Theory Was Wrong

Williams' treatment was based on the conclusion that modern upright posture "causes most low back problems,"[47] i.e. humans have back problems due to their erect posture, a posture that is "different from that of any of earth's other creatures."[48] Erect posture is the problem because humans are "physically ill-equipped to walk upright."[49] Part of the solution, he concluded, is to walk with the body tilted forward. To achieve this forward posture, the individual must force the lumbar spine *backward*, thereby changing the weight distribution to reduce the natural lordosis that results from upright walking.

---

[44]Sanders and McCormick, 1993, p. 440.
[45]Sanders and McCormick, 1993, p. 440
[46]Sanders and McCormick, 1993, p. 441.
[47]Williams, 1982, p. 13.
[48]Williams, 1982, p. 13.
[49]Williams, 1982, p. 13.

Williams also incorrectly concludes that walking upright is an "extremely difficult skill to master"–which is why "it takes a human child about three years to become an accomplished walker; whereas most other land animals become quite competent within the first few weeks following their birth."[50] He compares the human's "standing erect" problems to trying to "stand a soft drink bottle on its neck."[51] In fact, nervous system development requirements are a major reason for the length of time it takes for a baby to learn to walk.

The reason for back problems, Williams stressed, is because the sacral area of an infant is curved the wrong way, i.e. backward. which would be natural if humans walked on all fours. The vertebrae are similarly shaped in animals that walk on all fours. When walking upright, the feet still move forward, but the lower back is shifted *upward* instead of *forward,* producing the problematic lumbar hollow in the small of the lower back.

The theory concludes, as a result of the s-curve produced in the back from walking upright in a structure that evolved to walk on all fours, the distribution of weight is not even across the entire surface of each disc. Consequently, the uneven pressure forces the disc out towards the back side, producing a herniated disc. Back pain usually results from this intervertebral disc being forced out, putting pressure on certain nerves, especially on the nerve roots in the buttocks and leg that form the sciatica nerve.

The sciatica nerve usually causes the severe buttock and leg pain. In a child, the disc consists of a tough, gristly outer skin that surrounds the disc nucleus which contains a jelly-like substance.[52] With age, the nucleus substance becomes firmer until it is closer to the consistency of a wad of chewing gum.[53] Furthermore, the nucleus substance can dry up with age, making a herniated disc less a concern, but the disc can also shrink, causing other problems such as stenosis.

If the disc erupts in a child, the nucleus jelly-like substance exerts very little pressure on the nerve, and the disc soon is repaired by the body. In

---

[50]Williams, 1982, p. 13.
[51]Williams, 1982, p. 13.
[52]Osti and Moore, 1996.
[53]Williams, 1982, p. 17.

an adult, the nuclear substance continually presses on the nerve, and is repaired very slowly.[54] Consequently, according to the Williams' Darwin-centered theory, one should stay in bed, often for a considerable length of time, in order to allow the naturally defective spine time to repair itself.

In summary, Williams concludes that, forcing the body "to stand erect, severely deforms" the spine, "redistributing body weight to the back edges of the intervertebral discs in both the low back and neck ... . The fifth lumbar disc (and sometimes the fourth lumbar disc as well), ruptures," and the nuclear material "ruptures into the spinal canal causing pressure on the spinal nerves."[55] As noted, the solution Williams recommended was primarily to "always sit, stand, walk, and lie in a way that *reduces* the hollow [or curved lordosis] of the low back to a minimum."[56] This "first command" is repeated throughout his classic textbook.[57] Since he concludes that "the normal pressure of standing erect has caused one or both of the lowest two intervertebral discs to rupture," walking and even standing would exacerbate back problems.[58]

This theory often is repeated by prominent Darwinists as a major support of the view that humans evolved from ancestors that walked on all fours.[59] The problem is that this theory, although logical, is incorrect, which is why the application of the theory to solve back problems has produced an enormous amount of harm.[60] It now is recognized that the lumbar vertebrae curvature is critically important for back health, and the problem usually does not result from too much lumbar curvature as Williams' theory states, but from too *little* lumbar curvature. The natural lordosis helps to prevent disc rupture by putting pressure on the disc to keep it in place.

Williams provides numerous drawings and illustrations of various ways that one can eliminate the hollow in the back. He even recommends the sitting position we commonly call "slumping." This and many of the

---

[54]Urban, 1996.

[55]Williams, 1982, p. 18.

[56]Williams, 1982, p. 20, emphasis mine.

[57]Williams, 1982, for example, see pp. 24 and 35.

[58]Williams, 1982, p. 22.

[59]Miller, 1999; Pigliucci, 2001.

[60]Krehbeil, 1994.

other postures he recommended are the *opposite* of what is recommended today (and some of the positions and therapies he recommended are now recognized widely as major *causes* of back problems). Williams also claims that "lying flat on the back also increases low back pain in many individuals."[61] The recommendation now is to sleep on one's back.

Williams even states "two of the most popular forms of exercise—walking and jogging—are not recommended for low back pain sufferers, because most people, especially those in their middle and later years, walk or jog with a hollow in their low back."[62] In fact, walking, and similar exercise, are often the *best* therapy for back problems.[63] It is also recognized that proper back care can enable repair of a protruding and even a herniated disc "in a large number of patients."[64] In the words of Caplan, "the body has an amazing ability" to heal a protruding disc.[65]

The fact that weight lifters in good physical shape routinely lift several hundred pounds without problems also belies many of the assumptions behind the Williams argument. It also supports the conclusion that the reason back pain occurs in most cases is due to *improper* lifting, lack of exercise, inflammation, and/or weak back and abdominal muscles.[66] The strong back muscles help the discs to remain in their proper position, and can distribute the weight more equally, reducing enormously the likelihood of a disc slipping.

In reference to back surgery for slipped discs and chronic pain, Harvard Medical School trained doctor Andrew Weil wrote: "Two frequently recommended operations—laminectomy and spinal fusion—are far from guaranteed to eliminate chronic back pain. The vast majority of cases will respond to nonsurgical treatments, including exercise, rest, hypnosis, and stress reduction."[67]

---

[61]Williams, 1982, pp. 30, 35.

[62]Williams, 1982, p. 47.

[63]Goldmann and Horowitz, 2000.

[64]Nelson, et al., 1999; Gillan, et al., 1998; Harvey and Tanner, 1991; Hernandez, 1991; Murphy, 1977.

[65]Caplan, 1987, p. 66.

[66]Dunn, et al., 1976; Gallucci et al., 1995.

[67]Wallechinsky and Wallace, 1993, p. 117.

The Misguided Use of Back Problems as Proof of Darwinism

Back problems often are mentioned as proof of a "design flaw" in humans, and consequently are used by Darwinists as evidence for human evolution.[68] A common example is Elaine Morgan who, in her 1994 book *The Scars of Evolution*, repeats the now-refuted conclusion that "lower back trouble arises because the kink in the lumbar region of the spine makes it structurally weak and unstable. If extra strain is imposed on the back, the lowest vertebrae is liable to slip backward along the slope of the next one up. Such displacements may bring pressure on the nerves emerging from the spinal column, giving rise to pain which may be eased or cured by rest, but is liable to reoccur."[69]

Olshansky, et al., claims that "our bodies deteriorate because they were not designed for extended" upright posture.[70] Specifically, they mention our upright posture that was "adopted from a body plan that had mammals walking on all fours...our backbone has since adapted somewhat to the awkward change." They add these evolutionary "fixes do not ward off an array of problems that arise from our biped stance."[71] Price also concludes that one of the most persuasive evidences for evolution is what he calls the "problems of human design" flaws.[72] And one of the most often cited examples of "poor design" is the human back.[73]

This argument was summed up by a leading modern Darwinist who claimed: "If you were going to design a two-legged creature from scratch, rather than fashion one out of a four-legged creature, you'd do a better job than was done with us. (That's why so many of us have back trouble.)."[74] In the words of leading paleontologist Ian Tattersal, there is no better way to illustrate the poor design of humans than an evaluation of

our own much-vaunted species, *Homo sapiens*. As a result

---

[68]Tanner, 1981; Campbell, 1974; Cartmill, 1972.

[69]Morgan, 1994, p. 28.

[70]Olshansky, et al., 2001, pp. 52-53.

[71]Olshansky, et al., 2001, p. 51.

[72]Price, 1996, p. 256.

[73]Price, 1996, p. 257.

[74]Wright, 2001, p. 4.

of our upright, bipedal posture, we suffer a huge catalog of woes, including slipped discs, fallen arches, wrenched knees, hernias, and aching necks. No engineer, given the opportunity to design human beings from the ground up, would ever dream of confecting a jury-rigged body plan such as ours. But our innumerable afflictions can be understood as the consequence of adapting an ancestral four-legged body to a new, bipedal lifestyle.[75]

Professor Kenneth Miller adds that the human back disproves intelligent design because the

> many imperfections of the human backbone ...can hardly be attributed to intelligent design. They are easy to understand if we appreciate the fact that our upright posture is a recent evolutionary development. Evolution has taken a spinal column well adapted for horizontal, four-footed locomotion and pressed it into vertical, bipedal service. It works pretty well, but every now and then the stresses and strains of this new orientation are too much for the old structure. Intelligent design could have produced a trouble-free support for upright posture, but evolution was constrained by a structure that was already there. Chiropractors, of course, continue to reap the benefits.[76]

Last, a partnership between the University of Pittsburgh and Carnegie Museum of Natural History has been established to train medical students in Darwinism. One point the program stresses is humans have back problems because "our human ancestors evolved from walking on all fours to standing on their two hind legs... when our ancestors began walking only on their hind legs, it... led to chronic lower back pain..."[77]

This commonly used argument for Darwinism is actually a theological argument which is the reverse of Paley's watch argument—which

---

[75]Tattersal, 2002, p. 100.

[76]Miller, 1999, p. 101.

[77]Rossi, 2006.

tries to prove a Designer by demonstrating wise design in nature. The evolutionists' argument endeavors to demonstrate that God does not exist by documenting poor design in humans. Dembski notes that modern knowledge of anatomy has shown that Darwinism is without merit in this area.[78]

Dysteleological arguments such as these are not only misinformed but side-step the central issue.[79] Darwinists, in this case, attempt to shift the focus from the real issue, viz., how spine-bearing creatures could have evolved from spineless ones, a position for which no substantive evidence exists to why a designer designed a human spine in a certain way as opposed to another way.

Another problem for evolutionists is many animals that walk on all fours, especially inbred animals, also have some of the same back problems as humans.[80]

The Darwinist explanation is so common that one major popular back problem book had to refute it. It answered the question *"Is it true that simply walking upright is the main reason humans have so many back problems?"* by explaining this common misconception

> is based on neither evolutionary nor anatomical fact. Indeed, the spine is a marvel of evolution that allowed human beings to walk upright, thereby freeing their hands for more productive use. And walking on four legs is no assurance of a trouble-free back; certain breeds of dogs develop herniated discs. Also, degenerative arthritis of the spine has been found in birds and reptiles, among other species.[81]

What about cases in which a person exercises properly and otherwise takes care of their back and still has problems? Some evidence exists that a mutation that produces inborn defects may be responsible at least for some forms of back problems, specifically intervertebral disc problems

---

[78]Dembski, 1998.

[79]Woodmorappe, 1999.

[80]Frendin, et al., 1999; Sukhiani, et al., 1996; Gaschen, et al., 1995; Ness, 1994; Nakama, et al., 1993; Wright, 1989.

[81]Caplan, et al., 1988, p. 12.

and sciatica.[82] Another cause of back pain could be untreated forefoot varum defects, especially hyperpronation patterns, i. e., walking so that the body weight is not evenly distributed on the foot, causing more weight on one side compared to the other side.[83]

Conclusions

The Williams back problem theory is only one of many examples of evolutionary conclusions that, although seemingly convincing (and Krogman's article shows that they were very convincing even to creationists)—are wrong. Verna Wright, Co-director of Bioengineering at Leads University, calls the claim that upright posture is the major culprit for back problems in humans "nonsense."[84] The Williams therapeutic approach is appropriate only for certain abnormalities, such as stenosis abnormalities found in older people. Nonetheless, one who has experienced most of the common back problems usually can determine very quickly that many of the positions Williams recommends are those that *cause* back problems, and many of the ones he claims are "incorrect" actually help to solve the problem.

Conscientious long-term utilization of the McKenzie approach has been highly effective in alleviating back problems, whereas the Williams' approach has resulted in untold suffering, and possibly permanent damage, to millions of back patients. In corresponding with colleagues about this chapter, I was given many case histories. The following is typical:

> I have had lower back problems for the last 28 years, due to having lifted a heavy object improperly. I had many doctors tell me that I should walk with my shoulders forward as though I were walking uphill. I had three operations in the L5 area, including the injection of chymopapain. I was told to stay in bed for six weeks; I did it but it only made my back worse. I was given a list of exercises to do, based on the premise that my problem was due to evolution. Many years later, I went to

[82] Annunen et al., 1999.
[83] Rothbart, et al., 1995.
[84] Wright, 1989, p. 34.

a young neurologist who advised me to throw out all of those exercises and substitute one that involved strengthening the lower back muscles and also to take long walks. I followed his suggestions and have had very little back trouble since that time.[85]

How many millions of persons with back-pain suffered far longer, some even having surgery, which may do more harm than good, because of using a theory based on Darwinism, is very difficult to determine. No doubt the number worldwide is in the thousands, and likely the millions. Now surgery is recommended only in about two percent of all cases, usually for spinal stenosis or a combination of disc hernia, a corresponding pain syndrome, evidence of root irritation and failure to respond to six weeks of nonsurgical treatment.[86]

This is only one of many examples where Darwinism has misled research and has produced conclusions that have resulted in much harm, both to individuals and society as a whole. The solution to most back problems is proper posture and exercise that stretches and strengthens the four sets of muscles that provide support for the spine.[87]

---

[85]Lillo, 2001.
[86]Deyo, 1998, p. 52.
[87]Sobel and Klein, 2000, p. 18.

References

Annunen, S and 14 others. 1999. "An Allele of COL9A2 Associated with Intervertebral Disc Disease. *Science* 285(5426):409-412.

Baptist Medical System Back School Course. 1998.

Berger, E. 2000. "Late Postoperative Results in 1000 Work Related Lumbar Spine Conditions." *Surgical Neurology*, 54(2):101-106.

Campbell, Bernard G. 1974. *Human Evolution: An Introduction to Man's Adaptations. Second Edition*. Chicago, IL: Aldine Publishing Company.

Caplan, Deborah. 1987. *Back Trouble: A New Approach to Prevention and Recovery*. Gainesville, FL: Triad Publishing Company.

_____. 1988. *The Fit Back: Prevention and Recovery*. Alexandria, VA: Time-Life Books.

Cartmill, M. 1972. Chapter 4. "Arboreal Adaptations and the Origin of the Order Primate" in Russell Tuttle's (ed.) *The Functional and Evolutionary Biology of Primates*. Chicago, IL: Aldine-Atherton.

Clements, H. 1968. *What to do about a "Bad Back" and Disc Trouble*. London: W. Foulsham & Co. Ltd., 80 pp.

Delaney, P.M. and M.J. Hubka. 1999. "The Diagnostic Utility of McKenzie Clinical Assessment for Lower Back Pain." *Journal of Manipulative and Physiological Therapeutics*, 22(9):628-630.

Dembski, William. 1998. *The Design Inference*. New York: Cambridge University Press.

Deyo, Richard A. 1998. "Low-Back Pain." *Scientific American*, August, pp. 49-53.

Dunn, J.E., C.L. Johnson, and W. Cox. 1976. "Treatment of Lumbar Disks with Chymopapain." *Physical Therapy*, 56(4):399-402.

Foster, N.E., K.A. Thompson, G.D. Baxter, and J.M. Allen. 1999. "Management of Nonspecific Low Back Pain by Physiotherapists in Britain and Ireland. A Descriptive Questionnaire of Current Clinical Practice." *Spine*, 24(13):1332-1342.

Frendin, J., B. Funkquist, K. Hansson, M. Lonnemark and J. Carlsten. 1999. "Diagnostic Imaging of Foreign Body Reactions in Dogs with

Diffuse Back Pain." *Journal of Small Animal Practice*, 40(6):278-285, June.

Gallucci, M., A. Bozzao, B. Orlandi, R. Manetta, G. Brughitta, and L. Lupattelli. 1995. "Does Postcontras MR Enhancement in Lumbar Disk Herniation have Prognostic Value?" *Journal of Computer Assisted Tomography*, 19(1):34-38.

Gaschen, Lorrie; Johann Lang and Hansjuerg Haeni. 1995. "Intravertebral Disc Herniation (Schmorl's Node) in Five Dogs." *Veterinary Radiology & Ultrasound*, 36(6):509-516.

Gillan, M.G., J.C. Ross, I.P. McLean and R.W. Porter. 1998. "The Natural History of Trunk List, Its Associated Disability and the Influence of McKenzie Management." *European Spine Journal*, 7(6):480-483.

Goldmann, David and David Horowitz. 2000. *American College of Physicians Guide to Back Pain*. New York: Dorling Kindersley.

Gracovetsky, Serge A. "Function of the Spine from an Evolutionary Perspective" in *Volume 1. The Lumbar Spine. 2nd Edition.* Philadelphia, PA: W.B. Saunders Company, 1996, pp. 259-269.

Harvey, J. and S. Tanner. 1991. "Low Back Pain in Young Athletes. A Practical Approach." *Sports Medicine*, 12(6):394-406.

Hernandez, Conesa A. 1991. "The Current Treatment and Spontaneous Recovery of the Herniated Disk." *Anales De La Real Academia Nacional De Medicina*, 108(1):227-244.

Hides, J.A., G.A. Jull, and C.A. Richardson. 1996. "Multifidus Muscle Recovery is Not Automatic after Resolution of Acute, First-Episode Low Back Pain." *Spine*, 21(23):2763-2769.

_____. 2001. "Long-Term Effects of Specific Stabilizing Exercises for First-Episond Low Back Pain." *Spine*, 26(11):E243-8.

Ishmael, William K. and Howard B. Shorbe. 1960. *Care of the Back.* Philadelphia, PA: J.B. Lippincott Company, 24 pp.

Jacob, Gary and Robin McKenzie. 1996. *Spinal Therapeutics Based on Responses to Loading.* Chapter 12 in Liebenson.

Krames Communications. 1986. *Back Owner's Manual: A Guide to the Care of the Low Back.* Daly City, CA: Krames Communications,

Sixth Printing, 16 pp. Illustrated by Fran Milner.

Krogman, Wilton M. 1951. "The Scars of Human Evolution." *Scientific American*, 185(6):54- 57.

Lamb, Andrew. 2001. Letter dated Feb. 8.

Liebenson, Craig (Editor). 1996. *Rehabilitation of the Spine: A Practitioner's manual.* Baltimore: Williams and Wilkins.

Lillo, Clifford. 2001. Letter dated May 13.

Mayr, Ernst. 2001. *What Evolution Is.* New York: Basic Books.

McKenzie, Robin. 1988. *Treat Your Own Back.* Waikanae, New Zealand: Spinal Publications, 4th edition.

_____. 1990. *The Cervical and Thoracic Spine: Mechanical Diagnosis and Therapy.* Waikanae, New Zealand: Spinal Publications Ltd.

_____. 1996. *Mechanical Diagnosis and Therapy for Low Back Pain: Toward a Better Understanding.* Waikanae, New Zealand: Spinal Publications. Philadelphia: Saunders, 2nd edition.

_____ and Craig Kubey. 2000. *7 Steps to a Pain-Free Life: How to Rapidly Relieve Back and Neck Pain Using the McKenzie Method.* New York: Dutton.

Miller, Kenneth. 1999. *Finding Darwin's God.* New York: Cliff Street Books.

Mooney, Vert. 1988. Introduction in Robin McKenzie.

Morgan, Elaine. 1994. *The Scars of Evolution.* New York: Oxford University Press.

Murphy, R.W. 1977. "Nerve Roots and Spinal Nerves in Degenerative Disk Disease." *Clinical Orthopaedics and Related Research,* 129:46-60.

Murtagh, John; Clive Kenna, Chris Sorrell. 1997. *Back Pain and Spinal Manipulation: A Practical Guide.* Oxford: Butterworth-Heinemann, 2nd edition.

Nakama, Sanenori; Yasuho Taura, Hideki Tabaru and Masahiro Yasuda. 1993. "A Retrospective Study of Ventral Fenestration for Disk Diseases in Dogs." *Journal of Veterinary Medical Science,* 55(5):781-784.

Nelson, B.W., D.M. Carpenter, T.E. Dreisinger, M. Mitchel, C.E. Kelly,

J.A. Wegner. 1999. "Can Spinal Surgery be Prevented by Aggressive Strengthening Exercises? A Prospective Study of Cervical and Lumbar Patients." *Archives of Physical Medicine and Rehabilitation*, 80(1):20-25.

Ness, M.G. 1994. "Degenerative Lumbosacral Stenosis in the Dog: A Review of 30 Cases." Journal of Small Animal Practice, 35(4):185-190.

Nesse, Randolph and George C. Williams. 1994. *Why we get Sick: The New Science of Darwinian Medicine.* New York: Times Books.

Olshansky, S. Jay, Bruce Larnes and Robert Butler. 2001. "If Humans were Built to Last." *Scientific American*, 284(3):50-55.

Osti, Orso L. and Robert J. Moore. 1996. "Biochemistry and Histology of the Intervertebral Disc: Animal Models of Disc Degeneration" in *Volume 1. The Lumbar Spine. 2nd Edition*. Philadelphia, PA: W.B. Saunders Company, pp. 281-306.

Pigliucci, Massimo. 2001 "Design Yes, Intelligent No." *Skeptical Inquirer*, 25(5):34-42.

Price, Peter W. 1996. *Biological Evolution*. Fort Worth, TX: Saunders College Publishing.

Ratcliff, John D. 1975. *Your Body and How it Works*. New York: Delacorte Press.

Razmjou, H., J.F. Kramer, and R. Yamada. 2000. "Intertester Reliability of the McKenzie Evaluation in Assessing Patients with Mechanical Low-Back Pain." *Journal of Orthopaedic and Sports Physical Therapy*, 30(7):368-383; discussion 384-389.

Richardson, Carolyn, Gwendolen Jull, Paul Hodges, and Julie Hides. 1999. Foreword by Manohar M. Panjabi. *Therapeutic Exercise for Spinal Segmental Stabilization in Low Back Pain: Scientific Bases and Clinical Approach.* Edinburgh; Philadelphia: Churchill Livingstone.

Richardson, Carolyn A., Chris J. Snijders, Julie A. Hides, Leonie Damen, Martijn S. Pas, and Joop Storm. 2002. "The Relation Between the Transversus Abdominis Muscles, Sacroiliac Joint Mechanics, and Low Back Pain." *Spine*, 27(4):399-405.

Rothbart, Rrian, Kevin Hanson, Pul Liley and M. Kathleen Yerrat. 1995. "Resolving Chronic Low Back Pain: The Foot Connection." *American Journal of Pain Management.* 5(3)84-90.

Rossi, Lisa. 2006. Report issued by the University of Pittsburgh Medical Center dated February 28.

Sanders, Mark S. and Ernest J. McCormick. 1993. *Human Factors in Engineering and Design.* Seventh Edition. New York: McGraw Hill.

Samanta, Ash and Jo Beardsley. 1999. "Low Back Pain: Which is the Best Way Forward?" *British Medical Journal,* 318:1122-1123.

Shuman, David and George Staab. 1960. *Your Aching Back and what You can do about It.* New York: Gramercy Publishing Company.

Smail, Ronald. 1990. "Oh, My Aching Back!" *Creation Ex Nihilo,* 12(4): 20-21, Sept-Nov.

Sobel, Dava and Arthur Klein. 2000. *Backache: What Exercises Work.* New York. Barnes and Noble.

Stankovic, R. and O. Hohnell. 1995. "Conservative Treatment of Acute Low Back Pain. A 5- Year Follow-Up Study of Two Methods of Treatment." *Spine,* 20(4):469-472.

Sukhiani, H.R.; J.M. Parent, M.A.O. Atilola and D.L. Holmberg. 1996. "Intervertebral Disk Disease in Dogs with Signs of Back Pain Alone 25 Cases (1986-1993)." *Journal of the American Veterinary Medical Association,* 209(7):1275-1279.

Tanner, Nancy M. 1981. *On Becoming Human.* Cambridge: Cambridge University Press.

Tattersall, Ian. 1998. *Becoming Human: Evolution and Human Uniqueness.* New York: Harcourt Brace.

_____. 2002. "Science Versus Religion? No Contest." *Natural History,* 111(3):100.

Tuttle, Russel (editor). 1972. *The Functional and Evolutionary Biology of Primates.* Chicago, IL: Aldine Atherton.

Urban, Jill. 1996. "Biochemistry. Disc Biochemistry in Relation to Function" in *Volume 1. The Lumbar Spine. 2nd Edition.* Philadelphia, PA: W.B. Saunders Company, pp. 271-281.

Vertosick, Frank T. 2000. *Why We Hurt: The Natural History of Pain*. New York: Harcourt.

Wallenchinsky, David and Amy Wallace. 1993. *The Book of Lists: The '90s Edition*. Boston, MA: Little, Brown and Company.

Williams, Paul C. 1965. *The Lumbosacral Spine: Emphasizing Conservative Management*. New York: McGraw-Hill.

_____. 1982. *Low Back and Neck Pain: Causes and Conservative Treatment*. Springfield, IL: Charles C. Thomas.

Woodmorappe, J. 1999. "The Panda Thumbs Its Nose at the Dysteleological Arguments of the Atheist Stephen Jay Gould." *Technical Journal*, 13(1):45-48.

Wright, Robert. 2001. "The 'New' Creationism." *The Earthling*, April 16. https://slate.com/news-and-politics/2001/04/the-new-creationism.html.

Wright, Verna. 1989. "How Your Joints (And Other Points) Challenge Evolution." *Creation Ex Nihilo*, 11(3):32-34.

Yost, Graham (Editor). 1986. *American Family Health Institute: Back Problems*. Springhouse, PA: Springhouse Corporation.

Young, Emma. 2004. "TV is a Switch-Off for Back Muscles." *New Scientist*, 183(2462):8-9.

Chapter 3

Knees and Joints

Knee problems are one of the most common conditions brought to the attention of physicians today. Darwinists claim that a major reason for this fact is due to claims that the joint is poorly designed because the joint originally evolved. In fact, most all knee problems today are due to body abuse or overuse and disease, not poor design. The knee is the largest, most complex joint in the human body, but is also one of the most used (and abused) body joints. It is also a marvel of engineering and design. Furthermore, no evidence exists of knee evolution in the abundant fossil record.

The Problem

A common argument by Darwinists is that humans could not have been created, but rather must have evolved, because, they argue, we are poorly designed. One of the most common claims of putative poor design (or "dumb design" in the words of Niall Shanks) is the human knee joint. Shanks concludes that evolution from walking on all fours (as in apes) to the modern human bipedal locomotion is what "causes many problems from knee and ankle trouble to lower back pain."[1] Shanks provides an

[1]Shanks, 2004, p. 156.

Figure 3.1: Anatomy of the Knee

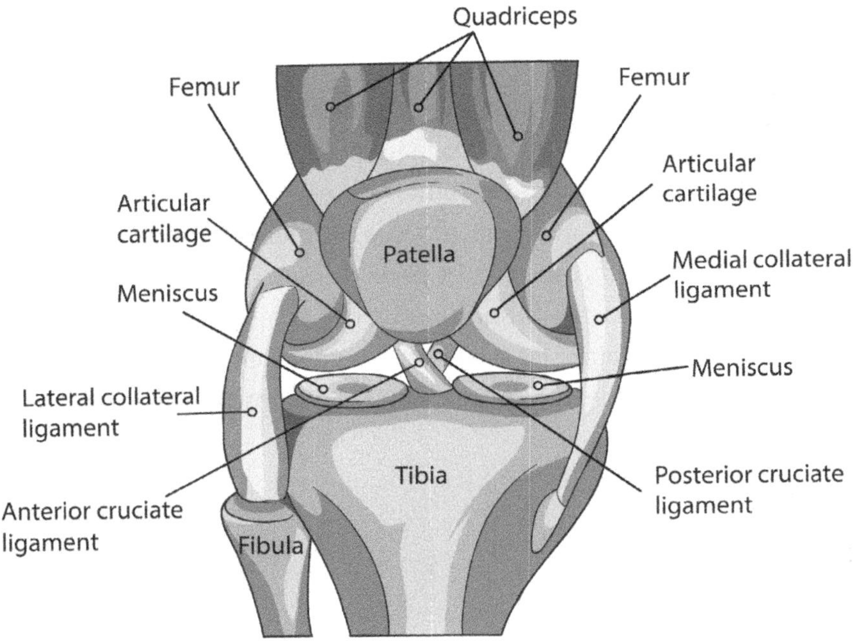

Image Credit: Viktoriia_P / Shutterstock.com

example of the "poor design" argument Darwinists use, which is actually a theological argument, as follows:

> To an evolutionary biologist, the appearance of poor design is evidence of the operation of a bungling, unintelligent trial-and-error evolutionary process that has resulted in suboptimal anatomical structures. Biologists point to these sorts of examples because they seem hard to account for if the intelligent design was due to an all-knowing, all-good, all-powerful designer, supernatural or otherwise. And this was precisely the sort of designer who has appeared in religious objections to evolution . . . if these defective structures were the result of design, then the designer must presumably have been drunk, stupid, or both![2]

Another, more colorful, example of the poor design claim is the following:

> If I ever catch up with the gonzo engineer who designed my knees, I'll sue Him (Her, It) for every penny. If any part of the body fits the description "not fit for purpose," it's knees and backs. He (She, It) was obviously having an off day. Goodness knows how it got past the quality assurance inspection.[3]

Professor David Barash concludes Intelligent Design is illogical because "the living world is shot through with imperfection," and the Creator is either incompetent or malevolent. A prime example he gives is the "ill-constructed knee joints that wear out." This, Barash concludes, proves that humans are "contingent, unplanned, entirely natural nature of natural selection. We are profoundly imperfect, cobbled together rather than designed. And in these imperfections reside some of the best arguments for our equally profound natural-ness."[4]

---

[2]Shanks, 2004, pp. 156-157.
[3]Kirk, 2005.
[4]Barash, 2005, p. 1.

Purported Evidence of Poor Knee Design

Claimed evidence for this claim include the fact that knee problems are responsible for over 18.3 million doctor visits annually.[5] Knee injuries rank second to lower back pain as the most common reason for outpatient physical therapy visits.[6] This is true for several reasons, including the knee is the largest and most complex joint in the human body, and one that must support the weight of almost the entire body—often 150 to 250 or more pounds.

When running, the forces can easily exceed 450 to 750 pounds at the contact points of each knee. Depending on the position of the knee joint, the area of contact can be as little as one cubic cm, meaning that a level of force as much as 750 pounds per cubic cm results.[7] The knee is designed so that the maximum distribution of forces results during any point of load bearing during knee motion.[8]

The knee is also more vulnerable to injury than other joints because it is one of the most mobile and flexible joints in the body, not because of design defects. To understand how well- designed the knee is, it is important to stress the extent of the use (and abuse) of this joint by the average human. By "the age of 85, even an average sedentary individual will easily have clocked 100,000 miles" and an active person over 200,000 miles, or almost 8 times around the world.[9]

If maintained by proper care, and not injured or abused such as in sports, the knee joint should last for over 200 million bends.[10] The more mobile a joint is, the less stable the joint is and, as a result, the more vulnerable it is to an injury. The knee joint is used over one million times per year and, as a result of all of these factors, it is one of the most injured joints in the human body. Actually, most all knee problems are due to

---

[5] 2004 American Academy of Orthopedic Surgeons data.
[6] Delitto, 1995, p. 1001.
[7] Müller, 1983.
[8] McLeod, 1994, p. 344.
[9] Simmons, 2004, p. 229.
[10] Simmons, 2004.

documented injury, abuse, or disease—not design defects.[11] Thus the "knee is one of the most used and abused joints" in the human body.[12]

The knee unit involves a complex set of bones, cartilage, muscles, tendons, ligaments, bursa, synovial membrane, sheathes, nerves, arteries, and veins all designed to work harmoniously together as a single, functioning unit.

The knee must achieve a balance between strength and the flexibility required to achieve the range of motion required for an active life, including in-line forward and twisting motions. Design changes to reduce the problem of misuse could compromise this balance:

> The compromise in the human knee between stability and flexibility is well illustrated in sports. The flexibility of our knees makes most athletics possible, but the price often is severe knee injuries. Most of these, particularly those incurred in playing basketball and football, result from violent pivoting of the body while the foot is firmly planted—a movement that twists the knee and tears the supporting ligaments that provide its stability.[13]

Artificial Knee Joints

Very few human inventions will last this long without major repairs. An artificial knee joint designed by the world's top scientists, and produced by the leading high tech corporations, typically, at most, lasts only around 20 years. The original usually lasts a lifetime, in spite of the fact that many of us abuse the joint. Part of the reason for this difference is that the knee is constructed out of bone, and no researcher has yet developed

> a material as well-suited for the body's needs as bone, which comprises only one-fifth of our body weight. In 1867 an engineer demonstrated that the arrangement of bone cells forms

---

[11]Segal and Jacob, 1989.
[12]Scott, 1994.
[13]Zihlman and Cramer, 1985, p. 10.

the lightest structure, made of least material, to support the body's weight. No one has successfully challenged his findings.[14]

Brand and Yancey also conclude that the design of bone produces a structure that

possesses incredible strength, enough to protect and support every other cell. Sometimes we press our bones together like a steel spring, as when a pole vaulter lands. Other times we nearly pull a bone apart, as when my arm lifts a heavy suitcase. In comparison, wood can withstand even less pulling tension, and could not possibly bear the compression forces that bone can. A wooden pole for the vaulter would quickly snap. Steel, which can absorb both forces well, is three times the weight of bone and would burden us down.[15]

Bone is "twice as tough as granite for withstanding compression forces, 4 times more resilient than concrete in standing up to stretching, about 5 times as light as steel."[16] After noting that "bone is an architect's dream," Werth notes the many advantages of using bone include it is

building material so malleable that it can be hammered into any shape, so versatile that when it's assembled into a light and durable framework it can execute and withstand complex mechanical movements, and so strong that it gives shape to and stiffens the whole human form without buckling. Not simply exquisite, as all great architecture must be, the edifice of the human skeleton is a perfect diagram of the lines of stress, tension, and compression involved in bearing the loaded structure—us—through a century or more of activity.[17]

---

[14]Brand and Yancey, 1980, p. 70.
[15]Brand and Yancey, 1980, p. 70.
[16]Werth, 2004, p. 76.
[17]Werth, 2004, p. 76.

The Knee is Optimally Designed

Professor Burgess concluded that the knee is a good example of optimal design. It contains "at least 16 critical characteristics, each requiring thousands of precise units of information" stored in the genome.[18] All of these structures must be present for the joint to achieve maximum function.[19] For example, the menisci are "vital components of the knee joint that assist in articular cartilage nutrition, shock absorption, and knee stability."[20] Its functional complexity has been demonstrated by anatomical studies of the knees of different species which have many structures in common implying that every

> anatomical component of this complex biomechanical system is needed for proper functioning of the whole system. Larson, in characterizing the knee as "the physiological joint," exquisitely described this concept. One may speculate about which characteristics of a four-bar linkage system have proved so advantageous to the design of the knee. Then one may ask which of these characteristics can be modified in the pathologically deficient knee to restore acceptable function.[21]

A four bar mechanism engineering structure is represented by four bars and four joints assembled so it can move in four directions.[22] Authors have struggled to explain how the knee functions in order to gain an "appreciation of the complex structure of the knee," a task "made easier by studying the analogous structure in the limbs of animals."[23] One problem in many mechanical mechanisms is finding a way to design a system so that the unit gives slightly, but does not collapse.[24] One major reason why knees function so well is because of the complex structure and chemical composition of bone. Specifically, the

---

[18]Burgess, 1999, p. 112.
[19]Burgess, 2000, pp. 17-25.
[20]Humphrey, 1998, p. 160.
[21]Dye, 1987, p. 983.
[22]see page 19, figure 10.4 in Burgess, 2000.
[23]Dye, 1987, p. 983.
[24]O'Connor and Goodfellow, 1978.

structural matrix of bone—a tight, interactive mix of protein and minerals—makes it a better building material than alloys and composites, but the true brilliance of its design is that it *lives*. The skeleton, like any living system, breaks down and renews itself continually. As the body grows to adulthood, it adapts its shape and proportions to match the demands of maturation. When bones break, they mend themselves.[25]

Furthermore, bones grow

outward from the middle of the shaft, the long bones that give the body its adult contours continue to grow until the age of 17 to 21. Brilliantly engineered to distribute force, the living skeleton not only bears the body's load and enables movement but also stores minerals, protects internal organs, and, in its spongy interiors, houses the main bloodworks.[26]

Conversely, if a part is too brittle, excess force will cause it to shatter. Armstrong notes that "the same problem has been encountered in the design of the human knee joint"—except a muscle is used instead of a spring:

The "force at the other end of the linkage" is commonly one's weight, with modifications according to a variety of situations. Now the knee joint is so designed, mainly by the shapes of parts that slide over each other, that the tension in the muscle is proportional to the "force at the other end" over quite a wide range of situations. Such an arrangement seems to be advantageous, in making it easy to adjust to a wide variety of situations. ... how such a design could possibly have evolved? Surely here is a very good engineering design, and, as usual, the design shows something of the skill of the Designer.[27]

---

[25] Werth, 2004, p. 78, italics in original.
[26] Werth, 2004, p. 80.
[27] Armstrong ,1971, p. 179.

The claim is also occasionally made that certain structures connected to the knee are unnecessary. For example, Müller[28] claimed certain ligaments were not needed; but other research indicates they are a necessary part of the knee design.[29] This is why Dye concluded, assuming Darwin years, that the

> general structural and functional similarity in the knees of diverse orders of animals implies that the knee is a profoundly adaptive biomechanical system that is unique among joints in tetrapods . . . the design of this joint has worked so well that it has persisted with little modification for more than 300 million years despite major modifications of functional demand.[30]

Knee Problem Types

Knee problems are classified into two major groups—mechanical and inflammatory. Mechanical problems usually result from injury—often a direct blow to the knee, or a rapid jerk, forcing the joint beyond the normal range of movement that the knee system is designed to sustain. This condition is common in certain sports, setting some people up for the potential of a lifetime of knee problems. Many sports knee injuries are caused by contact sports such as football. The knee is especially vulnerable because of the way the weight of the body impacts the knee. The knee cannot be designed to withstand a major foul/side impact and still achieve the needed everyday life flexibility.

Other knee problems result from—or are highly influenced by—poor lifestyle habits, including obesity, smoking, poor diet (such as diets low in calcium and vitamin D), and a sedentary lifestyle. A sedentary life causes the support muscle system to weaken and, as a result, the system is more likely to be injured when abused.

Most knee injuries are treated by allowing the joint to heal itself, ideally aided by the R.I.C.E. (Rest, use of Ice, Compression, and Elevation

---

[28]Müller, 1983.
[29]Wadia, et al., 2003.
[30]Dye, 1987, p. 982.

plus time).[31] The knee also acts like a fuse used in electrical circuits: knee pain or problems signal that the body is being overworked or abused. If the knee was able to take more abuse without pain, some persons would likely take their body beyond its limits and risk more serious and permanent damage to other body parts.

The second class of knee injury concerns include inflammatory problems that result from diseases, such as osteoarthritis and rheumatoid arthritis (an autoimmune disorder). These conditions are due to problems unrelated to knee design, such as genetic or body chemistry diseases.

Knee Evolution Claims

In spite of differences, the basic design of the knee joint is similar in higher animals and humans.[32] Even the knee of a chicken

> has several striking similarities to the human joint, including a bicondylar cam-shaped distal portion of the femur, relatively flat tibial plateaus, a patella, intra-articular cruciate ligaments, menisci, a broad and flat medial collateral ligament, and a more cylindrical lateral collateral ligament. The morphology of the knee in chickens also has differences, such as a femorofibular articulation and an extensor digitorum longus that originates on the lateral femoral condyle.[33]

The major part of the knee is bone; thus, the knee is well preserved in the fossil record of many animals. All extant "knees" in the fossil record are fully formed and developed, and no evidence exists of transitional forms.[34] Dye adds that the "complex functional morphologic characteristics of the knee are of ancient origin."[35] Research on fossils by Darwinists has determined that the "common ancestor of all living reptiles, birds, and mammals" called *Eryops* had a knee design very similar

---

[31]Mourad, 1991.
[32]Scott and Dye, 1987.
[33]Dye, 1987, p. 977.
[34]McLeod, 1994, p. 345.
[35]Dye, 2003, p. 19.

to the human knee.[36] Dye adds, *Eryops* had all of the "commonly shaped characteristics of the knees of most living tetrapods."[37] Furthermore, "the basic characteristics of the human knee are amazingly ancient in origin, dating back to 320 million [Darwin] years."[38] Dye notes that very early in history, according to the fossil record, the knee design was very similar to the design seen today and no evidence of knee evolution from a non-knee design. For example,

> the femur, tibia, and fibula were present and distinct, and the distal end of the femur already exhibited a bicondylar shape. The proximal part of the tibia was relatively flat and articulated with the preaxial condyle of the femur. The post-axial condyle of the femur (lateral femoral-condyle analogue) articulated with the proximal part of the fibula. This fibular articulation with the femur remains a characteristic of knees in reptiles, birds, and some primitive mammals.[39]

Interestingly, the "study of comparative anatomy demonstrates the similarities of [knee] design among the tetrapods."[40] This indicates to Darwinists a common knee origin, but an objective evaluation indicates the similarity is due to its effective design, and highlights the functional effectiveness of the joint's design. As Hosea, et al., conclude, the fact that the knee "has functioned with little alteration for more than 300 million years despite major functional demand changes" demonstrates the high effectiveness level of knee design. [41] The details of this similarity include the fact that the functional dynamics of all of the knees that have been dissected to date are similar to the human knee design:

> Each knee has a complex rolling and gliding motion of the femur on the tibia, with the point of contact on the femur moving posteriorly on the tibia with flexion, like a four-bar

---

[36]Hosea, et al., 1994, p. 3.
[37]Dye, 1987, p. 978.
[38]Hosea, et al., 1994, p. 3.
[39]Dye, 1987, p. 977.
[40]Hosea, et al., 1994, p. 3; see also Dye, 1987, p. 976, and Hinchliffe and Johnson, 1980.
[41]Hosea, et al., 1994, p. 3.

linkage system.... The similarity of the asymmetrical design of the medial collateral ligament is also an unexpected finding. ... in all of the species that have been dissected to date, the medial collateral ligament was found to be flat and broad and to have a tibial insertion well distal to the joint line.[42]

Although no evidence exists for knee evolution, and fully formed knees appear at the beginning of the tetrapod fossil record, much variety exists based on the basic knee design, both then and today. All of the "complex set of attributes that we associate with the human knee" are "extremely ancient in origin" and exited from the beginning.[43] The design of the human knee is basically the same in all mammal knees, yet is distinctly different in several ways. Dye notes that no ideal animal model is known for the human knee.[44] A major difference is, only the human knee is designed to lock easily in the extension (straight leg) position to allow maintaining comfortable vertical posture for a sustained period of time.[45]

This design feature is one reason why humans are able to easily walk and run upright. Apes' knees cannot lock, and must be continually loaded in the flexion (bent leg) position, requiring a great amount of muscle use resulting in rapid tiring. Try standing with your knee's slightly bent for ten minutes or so—you soon will note this position is extremely tiring. The "locking" mechanism occurs only in extension. As soon as motion/flexion occurs by action of the popliteal muscle, tibial rotation occurs to "unlock" the knee.

For this reason, apes are generally quadrupedal (four-legged), and it is extremely difficult for them to maintain a vertical posture for any length of time. Humans, in contrast, are biped (two-legged) and cannot efficiently walk on all fours as can apes. The only way apes can stand upright is by awkwardly bending their ankle, knee, and hip joints. Such a distorted posture means that apes can stay in a vertical position for only short periods of time and distances. In contrast, an able-bodied fit human

---

[42]Dye, 1987, p. 979.
[43]Dye, 1987, p. 982.
[44]Dye, 1987.
[45]McLeod, 1994, p. 343.

can stand for hours, or run for many miles, without similar difficulty.[46]

Improving Knee Design

Olshansky, et al., to support their claim that the creation worldview is wrong, argue that the body is poorly designed to prove that we evolved by an impersonal, non-theistic process. A major example they use in an attempt to prove this thesis is the human knee joint. They even argue that the human knee should be completely redesigned to include a thicker cartilage pad to allow the human knee to "bend backward" as do apes' knee joints. They also advocate removal of the knee cap (the patella). Olshansky, et al., admit that their design also has its problems, such as "the absence of a locking mechanism would make it hard to stand for very long, so further modifications would be necessary."[47]

The patella (which Olshansky eliminated) is a free floating sesamoid bone (a bone that is independent rather than articulating with another bone), which is also critically important. Its functions include serving as a lever to allow much greater leg strength, and its loss would seriously handicap coordinated body movement. The massive quadriceps femoris muscle is connected to a large tendon that passes over the patella to attach to the tibia. The patella also helps to protect the underlying bones and tissue in the knee joint area. Olshansky's et al,. "improved" knee is clearly an inferior design! The only way to test its design is to surgically alter knees of patients to determine if the design claimed to be superior is, in fact, actually superior.

To improve artificial joint replacement and bracing systems, scientists need to more closely copy the original design and consider the complexities and functional morphologic features of the healthy human knee.[48] The human knee is an excellent example of why a Harvard professor of medicine said about the human body: "Learning how this awe-inspiring and remarkably intricate piece of machinery is assembled—how it works,

---

[46]Burgess, 1999.
[47]Olshansky, et al., 2003, p. 97.
[48]Dye, 2003, p. 19.

Figure 3.2: X-Ray of Knee Replacement

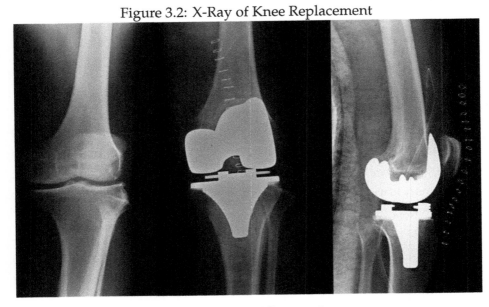

Image Credit: Jarva Jar / Shutterstock.com

its control and communication systems, and its central programming—occupies the full attention... of the Harvard Medical School" students.[49]

Summary

Various claims about the putative poor design of the knee are erroneous. The knee is an excellent example of evidence for design. The fact, as one evolutionist admitted, is "the knee joint is among the greatest of nature's inventions, 'a 360-million-year-old structure beautifully designed to do its job of transferring load between limps.'"[50] Most common knee problems are not due to poor design, but disease, abuse, injury, wear and tear, natural calamities, and aging. The knee is actually a masterpiece of design and, in the absence of these factors, the knee will normally function as a very resilient structure that can be expected to give its owner over eight

[49]Moore, 1995, p. 3.
[50]Ackerman, 2006, p. 137.

decades of good maintenance-free service. The knee allows humans to walk upright, and our bipedal locomotion is

> a whole-body activity that involves a complex set of interre-lated behaviors, including carrying and throwing, rather than simply as a way of moving from one place to another. Our kind of two-legged locomotion allows a wide variety of motor patterns: it is a "doing" system. For example, no other pri-mate can throw with precision, or walk long distances while carrying objects in its arms and hands.[51]

The conclusion that, if God designed the knee, He would have de-signed it very differently is actually a theological, not a scientific argu-ment, as are all poor design claims. This argument claims to second-guess the thoughts of the Creator. Theological arguments require considering that the design of all body organs have been compromised by the Fall. On the other hand, if "our knees seem flawed and liable to give out, it's because we ask an awful lot of them."[52]

The fossil record also shows that, although several distinct designs of knees exist, no evidence exists for knee evolution. This conclusion is not due to a poor fossil record, because an excellent fossil record exists for the reason that bone is preserved better than most all other body structures—only teeth are usually better preserved.

---

[51]Zihlman and Cramer, 1985, p. 67.
[52]Komaroff, 2006, p. 1.

References

Ackerman, Jennifer. 2006. "The Downside of Upright". *National Geographic*. July. pp. 127-145.

Armstrong, Harold. 1971. "Comments on Scientific News and Views: Springs and Knees—Design." *Creation Research Quarterly*, 8(3):197.

Barash, David P. 2005. "Does God Have Back Problems Too? The Illogic Behind 'Intelligent Design.'" *Los Angeles Times*. June 27.

Brand, Paul and Philip Yancey. 1980. *Fearfully & Wonderfully Made: A Surgeon Looks at the Human & Spiritual Body*. Grand Rapids, MI: Zondervan.

Burgess, Stuart C. 1999. "Critical Characteristics and the Irreducible Knee Joint." *Technical Journal*, 13(2):112-117.

_____. 2000. *Hallmarks of Design*. Surrey: Day One Publications. Chapter 1, pp. 10-34.

Dye, Scott F. 1987. "An Evolutionary Perspective of the Knee." *Journal of Bone and Joint Surgery*, 69(7):976-983.

_____. 2003. "Functional Morphologic Features of the Human Knee: An Evolutionary Perspective." *Clinical Orthopaedics and Related Research*, 410:19-24, May.

Delitto, Anthony. 1995. "Lower Extremity: Knee." In Myers' *Saunders Manual of Physical Therapy Practice*, Chapter 34, pp. 1001-1029. Philadelphia: W.B. Saunders Company.

Hinchliffe, J.R. and D.R. Johnson. 1980. *The Development of the Vertebrate Limb*. Oxford: Clarendon Press.

Hosea, Timothy M.; Alfred J. Tria, and Jeffrey R. Bechler. 1994. "Chapter 1: Embryology of the Knee," pp. 3-13, in *The Knee* (W. Norman Scott, editor).

Humphrey, Ellen Cook (editor). 1998. *Programs in Physical Therapy Northwestern Medical School Clinical Orthopedics Clinical Conditions Papers—1998*.

Kirk, Steve. 2005. "Unintelligent Design." *New Scientist*, July 23, p. 20.

Komaroff, Anthony. 2006. Knees in Need" *Harvard Health Letter*. 31(7):1-3.

McLeod, Kevin. 1994. "Knee Design: Implications for Creation vs. Evolution." *Third International Conference on Creation*, pp. 343-349.

Moore, Francis. 1995. *A Miracle and a Privilege: Recounting a Half Century of Surgical Advance*. Washington, D.C.: Joseph Henry Press.

Mourad, Leona. 1991. *Orthopedic Disorders*. St. Louis, MO: Mosby.

Müller, Werner. 1983. *The Knee: Form, Function and Ligament Reconstruction*. New York: Springer.

Myers, Rose Sgarlat. 1995. *Saunders Manual of Physical Therapy Practice*. Philadelphia: W.B. Saunders Company.

O'Connor, J. and J. Goodfellow. 1978. "The Mechanics of the Knee and Prosthesis Design." *Journal of Bone and Joint Surgery*, 60B:358-369.

Olshansky, S. Jay, Bruce Carnes, and Robert Butler. 2003. "If Humans were Built to Last." *Scientific American* (Special Edition), 13(2):94-100.

Roberts, T.D.M. 1971. "Standing with a Bent Knee." *Nature*, 230(5295):499-501.

Scott, F. and M.D. Dye. 1987. "An Evolutionary Perspective of the Knee." *Journal of Bone and Joint Surgery*, 69A:976-983.

Scott, W. Norman (Editor). 1994. *The Knee*. St. Louis, MO: Mosby.

Segal, Philippe and Marcel Jacob. 1989. *The Knee*. London: Wolfe.

Shanks, Niall. 2004. *God, the Devil, and Darwin*. New York: Oxford University Press.

Simmons, Geoffrey. 2004. *What Darwin Didn't Know*. Eugene, OR: Harvest House.

Wadia, F.D.; M. Pimple, S.M. Gajjar, A.D. Narvekar. 2003. "An Atomic Study of the Popliteofibular Ligament." *International Orthopaedics*, 27(3):172-174.

Werth, Barry. 2004. *The Marvel of the Human Body, Revealed*. New York: Doubleday.

Wilson, Stephen A, Vincent J. Vigorita, and W. Norman Scott. "Chapter 2: Anatomy," pp. 15- 54, in *The Knee* (W. Norman Scott, Editor).

Zihlman, Adrienne and Douglas Cramer. 1985. "Human Locomotion." *Natural History* 85(1):65-68.

Chapter 4

The Hands and Feet

Orthodox Darwinism postulates modern humans evolved from a hypothetical apelike common ancestor. Consequently, theories of human foot and hand evolution are based on comparisons of hypothetical pre-human feet and hands using modern chimp feet and hands as the model.[1] These comparisons break down when examined in more detail. Furthermore, claims of poor design of the feet are also very problematic when examined carefully.

The Overly Complicated Human Hands and Feet Claim

The consensus of a panel of anthropologists at the annual meeting of the American Association of the Advancement of Science (AAAS) was that although "humans are the most successful primates on the planet," our "bodies would not win any awards for good design." To give one of many examples, they asserted

> how poor a job evolution has done sculpting the human form
> . . . . Using props and examples from the fossil record, the
> scientists showed how the very adaptations that have made

---

[1]Harcourt-Smith, W.E.H. and Aiello, L.C., Fossils, feet and the evolution of human bipedal Locomotion, *J. Anatomy* 204:403–416, 2004.

Figure 4.1: Human Foot Muscle, Bone, Tendon and Nerve Structures

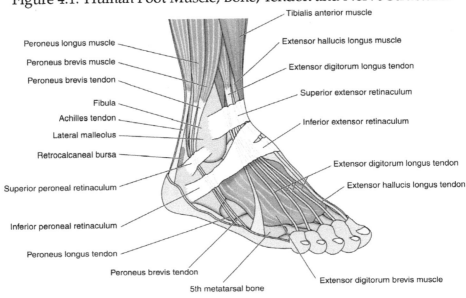

Image Credit: Blamb / Shutterstock.com

humans so successful—such as upright walking and our big, complex brains—have been the result of constant remodeling of an ancient ape body plan that was originally used for life in the trees.[2]

Boston University anthropologist Jeremy DeSilva explained that human feet have too "many bones because our ape-like ancestors needed flexible feet to grasp branches" of trees.[3] DeSilva added "modern foot anatomy isn't what you'd design from scratch," because "Evolution works with duct tape and paper clips." He then showed his AAAS audience a human foot cast with 26 bones, claiming using 26 bones is poor design because, although our purported ape-like ancestors required flexible feet to grasp branches,

as they moved out of the trees and began walking upright on the ground in the past 5 million years or so, the foot had to

---

[2]Gibbons, 2013.
[3]Gibbons, 2013.

become more stable, and bit by bit, the big toe, which was no longer opposable, aligned itself with the other toes and our ancestors developed an arch to work as a shock absorber.[4]

The modifications were made to allow the human foot to remain comparatively rigid, but in order for this system to work, a "lot of Band Aids" were required. The result was

> our foot still has a lot of room to twist inwards and outwards, and our arches collapse. This results in: ankle sprains, plantar fasciitis, Achilles tendonitis, shin splints, and broken ankles. These are not modern problems, due to stiletto heels; Fossils show broken ankles that have healed as far back as 3 million years ago.[5]

DeSilva claimed a much "better design for upright walking and running would be a foot and ankle like an ostrich. An ostrich's ankle and lower leg bones are fused into a single structure, which puts a kick into their step—and their foot has only two toes that aid in running."[6] He added the reason for the poor design of the human foot is because "ostriches trace their upright locomotion back 230 million years to the age of dinosaurs, while our ancestors walked upright just 5 million years ago."

The same argument was given for the putative poor design of the human hand and wrist. As Professor Lents writes, referring to the human wrist: "There are eight different bones in the wrist... it is way more complicated than it needs to be. No sane engineer would design a joint with so many moving parts."[7] He adds that this "small area that is just the wrist itself has eight fully formed and distinct bones tucked in there like a pile of rocks—which is about how useful they are to anyone."

Lents added that the "human ankle suffers from the same clutter of bones that we find in the wrist. The ankle contains seven bones, most of them pointless." He admits that the "ankle certainly has more to deal

---

[4]Gibbons, 2013
[5]Gibbons, 2013
[6]http://www.sciencemag.org/news/2013/02/human-evolution-gain-came-pain.
[7]Lents, 2018, p. 28.

with than the wrist does, given that it is constantly bearing weight and is central to the locomotion of the entire body," but argues that "they would function better as a single fused structure."[8] This claim is even made in the popular science press.[9] In short:

> Evolution doesn't "design" anything, says anthropologist Matt Cartmill of Boston University. . . . It works slowly on the genes and traits it has at hand, to jerry-rig animals' and humans body plans to changing habitats and demands. "Evolution doesn't act to yield perfection," he says. "It acts to yield function."[10]

A Response to these Claims of Poor Design

Anthropomorphizing evolution as if it acts like an intelligent sentient being is the first problem. Secondly, the evolution claim is problematic because the "human foot is considered one of our most distinctive morphological and functional features" by those who specifically study the system.[11] Some anthropologists have even concluded that the "feature that differentiates Hominids from other primates is not large brain size, but the set of characters associated with erect bipedal posture and a striding gait."[12]

One example is our distinctly human big toe, which is a mechanical device that functions as a powerful lever to rise the entire body. Conversely, the ape big toe does not function as a toe, but rather as a thumb.[13] To convert the foot of a gorilla into that of a man requires numerous major structural changes, including rotation of the gorilla toe so it rests flat on the ground like the other toes instead of *facing* the other toes like a thumb. Last, the bones must also be shortened and the socket and other parts drastically altered so that it lies parallel to the other toes.[14]

---

[8]Lents, 2018. pp. 28-29.
[9]Greenwood, 2014.
[10]Gibbons, 2013.
[11]Bates, et al., 2013.
[12]Carrier, 1984, p. 483; see also Wayman, 2012.
[13]Stern, 1928, p. 5.
[14]Stern, 1928, p. 5.

The flexibility Darwinists complain about that is produced by the 26 bones and 33 joints is required to allow us to achieve the many tasks our active life as humans requires. The foot's 19 muscles and 57 ligaments, plus the nerves required to coordinate the foots' entire system's functions exist for a reason. The bones in your hands and feet for very good reasons. The foot is one of the few parts of our anatomy that can compete with the hand for its enormous versatility due to its sheer complexity.

Most of humanity has walked barefoot since our creation, requiring a foot design that is able to traverse uneven ground containing dirt clumps, pebbles and other detritus. This trait requires a very flexible foot, not a flat, stiff foot. The skin on the bottom of the foot is both thick and tough. Under the foot's heel lies a pad of specialized fat that is packaged like bubble wrap to effectively absorb shock and spread the bodies' weight load to a larger area. Some podiatrists and others argue that shoes have contributed to some of our foot problems, adding that barefoot walking and running may have some advantages for foot health.[15]

If any changes are desirable for most of humanity, one may be a *more flexible* foot. However, this would require a trade-off with other characteristics, such as reduced strength or resistance to injury. In short, our feet are strong enough to respond to some of the greatest forces the body experiences, whilst at the same time they are capable of complex fine movements as exquisite as a ballerina's pirouette, or even to write letters or paint pictures as required by persons who have lost their arms due to birth defects or accidents.

Proprioception is the sense that tells us where the various parts of our body are located in space. It allows us to touch our nose with any finger while our eyes are shut. Sense organs in our muscles and joints tell the brain when, and exactly where, our limbs, feet and hands are located and the tension our muscles require to move limbs to some specific location. This continuous feedback system between muscle and brain detects minute changes that allows us to do everything from playing the piano, to catching a baseball, to balancing ourselves as we walk and stand.

That human feet are not designed for grasping or climbing trees is

---

[15]Jenkins, and Cauthon, 2011.

obvious from comparing human hands with ape hands:

> Human hands are unique compared to the rest of the animal kingdom, as are our feet. The combination of an opposable thumb (the thumb being able to touch the tip of any of the other fingers) and our being able to rotate the hand freely (palm up or palm down, shared with primates) gives us an ability to manipulate objects and tools like no other creature alive, even if they felt so inclined.[16]

The design that allows hands to achieve the wide range of activities described above requires many trade-offs to optimize the entire design. Consequently, abuse from a wide variety of sources can damage the system. For example, when running one can twist an ankle, a painful event that may take some time to heal. An ankle sprain can occur when the foot suddenly twists or rolls when running, forcing the ankle joint out of its normal position. This puts a great deal of force on the ankle at an angle that it was not designed to deal with, causing one or more ligaments around the ankle to stretch beyond its normal length or even tear.

A major contributor to the problem of ankle damage is a systemic weakness and lack of good flexibility resulting from conditions such as unsatisfactory shoes or lack of exercise. To reduce the problem it is recommended that someone who is out of shape should perform strengthening exercises, wear sturdy, quality footwear, pay attention to surfaces that one is walking on, slow or stop activities when feeling fatigued, and always warm up before exercising. Women should avoid high heels and both sexes should avoid poor fitting shoes.

The Enormous Force Absorbed by the Foot

A force of up to three times our body weight is exerted on the human foot while running. A person weighing 200 pounds will experience forces of up to seven times their body weight or almost 600 pounds Conversely,

---

[16]Vetter, 1990.

a healthy younger person in good physical shape may be able to avoid injury even in the case of twisting an ankle, although some swelling or bruising may occur as a result of this 1,400 pounds force.

Tendons, cartilage, and blood vessels may also be damaged due to a sprain, causing swelling, pain or discomfort when weight is placed on the affected area. Ankle sprains can happen to anyone at any age but older persons are far more susceptible. Participating in sports, walking on uneven surfaces, or even wearing inappropriate footwear can all contribute to this type of injury.

At the core of human foot design is the arch. The metacarpal bones create the space required for a tough web of muscles and ligaments that effectively absorb much of the forces created when walking or running.[17] To support this weight designed in the system is an ingenious "windlass" design. It allows the bones' joints to relax when the foot hits the ground and then, when moving forward, it tightens to help propel the person forward. This is one reason for a design utilizing a large number of bones. At least four arches, using a keystone similar to the Roman Arch design, is required to achieve these conflicting demands. This system effectively provides the flexible design that is required for the foot to land, and then initiates the rigid propulsion stance to take the next step.

Add Illustration About Here

The arch functions as a spring; storing and releasing energy when you push off using your toes.[18] Today in the West we usually run wearing cushioned shoes, and this may be why we tend to land heel first. This practice risks damaging the joints because the impact shock travels up through our legs. Conversely, barefoot runners, when their foot hits the ground, tend to make simultaneous contact with the balls of their feet and the foot arch, a landing that safely dissipates the impact energy.[19]

---

[17]Carrier, 1984.
[18]Zimmer, 1995.
[19]Jenkins, and Cauthon, 2011.

The Design of the Hand

In many ways it is because of our foot design that we have such extraordinary hands. The ability to walk upright allows humans to efficiently cover great distances. But it also frees up our hands to use their unique anatomy and capabilities. Even evolutionists recognize the excellent design of the human hands and feet. One wrote: "Both the human hand and foot represent a triumph of complex engineering, exquisitely evolved to perform a range of tasks... The hand is one of the most complex and beautiful pieces of natural engineering in the human body."[20] Most of its movements are controlled by muscles located in the forearm connected to the finger bones via the strong long tendons that pass through a flexible wrist.

This remote musculature design provides the fingers with both movement and strength that would be impossible if all of their muscles were attached directly to each finger. Some describe the hand as a boney puppet controlled by the ligaments located in the forearm. The impressive strength of a mountain climbers' hands is one example. As a result of habitual use and training a single finger can support your entire body weight. At the other extreme is the control required for a pianist that is achieved by the hands' intrinsic muscles, namely the lumbricals.

As noted the muscles are not directly attached to bones. They attach to tendons to allow fine, subtle movement. Some of these muscles specifically control the thumb and little finger. The little finger's importance is second only to the thumb. Conversely, the finger you can lose with minimum inconvenience is the index finger. It can be included, or excluded, from most every activity that we normally use our hands for.

Even fingernails play a crucial part in the achievements of the hand. Pressing against their rigid structure allows us to be able to judge how firmly to hold on to something so as not to drop it. In short, the "hand is one of the most complex and beautiful pieces of natural engineering in the human body. It gives us a powerful grip but also allows us to manipulate small objects with great precision. This versatility sets us apart from every

---

[20]McGavin, 2014.

Figure 4.2: Muscles of the Hand

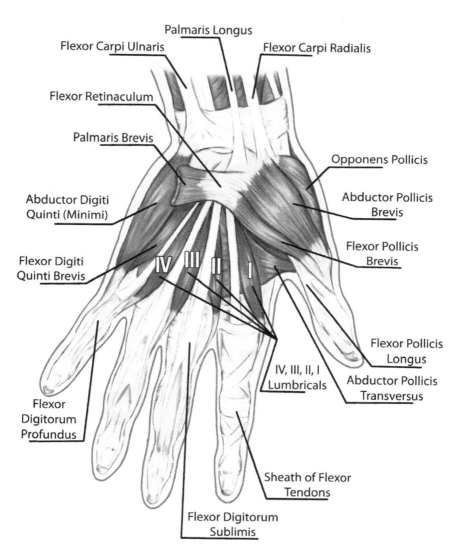

other creature on the planet."[21] Dr. McGavin, although an evolutionist, concluded that human

> hands and feet are biomechanical marvels. More than any other piece of anatomy, they are what have made us such a successful species. They've allowed us to walk out of Africa to colonize the globe and master the natural world. I will never look at my hands and feet in the same way again. [22]

A major evidence of the importance of the many bones in the hand and feet occurs when these bones become fused due to arthritis, birth defects, or injury. Often the hands and feet lose flexibility, causing a major impediment in the ability of the victim to carry out daily tasks. The many bones in the hands and feet are designed for a very good reason, and to argue the opposite is irresponsible.[23] Despite claims of poor designs in biology, evolutionists consistently fail to outline better designs and test them, much less somehow implement the "improved" design, They typically lack real expertise, such as the panel of anthropologists - not engineers or other experts in mechanical design - opining on the design of the foot.

---

[21]McGavin, 2014.
[22]McGavin, 2014.
[23]Singh, et al., 2003.

Figure 4.3: Bones of the Hand

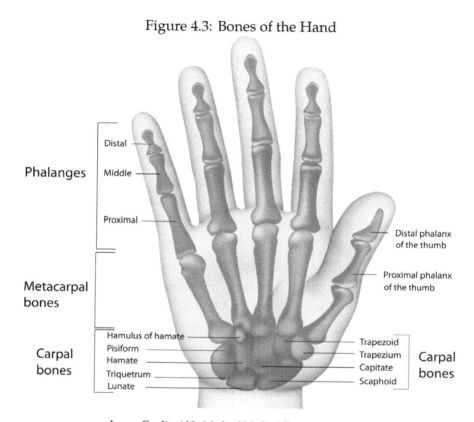

Image Credit: Alila Medical Media / Shutterstock.com

References

Bates, K. T. et al., 2013. The Evolution of Compliance in the human Lateral Mid-Foot. *Proceedings of Biological Science.* 2013 Aug 21;280(1769):1-8.

Carrier, David, 1984. The Energetic Paradox of Human Running and Hominid Evolution. *Current Anthropology.* 25(4):483-495.

Gibbons, Ann. 2013. Human Evolution: Gain Came With Pain. https://www.sciencemag.org/news/2013/02/human-evolution-gain-came-pain.

Greenwood, Veronique. 2014. You're not highly evolved: It's not just the appendix. *Popular Science.* May.

Jenkins, David W., and David J. Cauthon. 2011. Barefoot Running Claims and Controversies. A Review of the Literature. *Journal of the American Podiatric Medical Association.* 101(3):231-246.

Lents, Nathan H. 2018. *Human Errors: A Panorama of Our Glitches, from Pointless Bones to Broken Genes.* Boston: Houghton Mifflin Harcourt.

McGavin, George. 2014. The incredible human hand and foot. http://www.bbc.com/news/science-environment-26224631.

Singh, P., A. Tuli, R. Choudhry, and A. Mangal, 2003. Intercarpal Fusion — A Review. *Journal of the Anatomy Society of India.* 52(2):183-188.

Stern, Bernhard. 1928. How Man Differs From the Ape. *Evolution.* 1(3):4.

Vetter, Joachim. 1990. Hands and feet—uniquely human, right from the start! *Creation* 13(1):16–17 December.

Zimmer, Carl. 1995. Best Foot Forward. Toes Help Humans Shift into Gear, Study Says. *Toledo Blade.* March 20, p. 23.

Part II

Design of the Reproductive System

Chapter 5

The Birth Canal

The poor design claim is a favorite ploy of Darwinists to argue against the Intelligent Design of humans. One of the most common examples of poor design the problem that childbirth presents for humans. The claim evolutionists make is that the birth mechanism was well-designed in our evolutionary ancestors, but the comparatively rapid enlargement of the human brain as we evolved created problems. They claim the main problem is the female human birth canal diameter did not correspondingly evolve larger to accommodate the evolution of a larger brain.

As one writer claimed, the "pelvis in the human female is an example of perfectly decent mammalian design rendered problematic by evolution. . . . . on account of our wonderfully large brains, human babies have huge heads, which compounds the problem of the tilted pelvis."[1] The claim of Darwinists is that the human female

> birth canal passes through the pelvis. The prenatal skull will deform to a surprising extent. However, if the baby's head is significantly larger than the pelvic opening, the baby cannot be born naturally. Prior to the development of modern surgery (caesarean section), such a complication would lead to the death of the mother, the baby, or both.[2]

[1] http://www.independent.co.uk/life-style/health-and-families/features/what-are-the-greatest-design-flaws-of-the-human-body-10279557.html.
[2] https://en.wikipedia.org/wiki/Argument_from_poor_design.

63

Another evolutionary argument is that the human pelvis is narrow for the reason that natural selection selected this design because it is much more efficient for both walking and, especially, running. Thus, evolution did not favor a wider birth canal when humans were evolving to walk upright, rather it favored the narrow design for its efficiency in escaping enemies. One study contradicted this claim, concluding that a *broader* pelvis "is substantially more efficient for locomotion than a narrower one."[3]

Abby Hafer, an Oxford trained biology Ph.D. who now trains nurses, viewed this argument against creationism of such importance that she devoted an entire chapter on the poor design of the female birth canal in her new book. She reasoned it was bad design because when the nine-month gestation has passed the baby has grown in size resulting in the problem that its "head has to fit through a circle of bone that is smaller than the head." [4] She concluded "One would think that a benevolent Creator would not make child-birth into such a problem in the first place. . . if only we *had* been designed rather than evolved." She concluded, therefore, that humans "evolved rather than being designed."[5]

The implications are, if "we claim that organisms and their parts have been specifically designed by God, we have to account for the incompetent design of the human. . . birth canal."[6]

Hafer defended this common poor design claim is based on the conclusion that humans and chimps have a common ancestor. As a result, the claim argues, the human head at birth was often too large to pass easily through the birth canal opening.[7] Consequently, she concluded, the size problem causes major birth problems for humans. This theory, called the *obstetrical dilemma*, is the current dominate evolutionary theory.[8]

---

[3]Vidal-Cordasco, M. ; et al. 2017 p. 609-622. November.
[4]Hafer, 2016. Chapter 11 Bad Design—The Birth Canal. pp. 45-52.
[5]Hafer, 2016. pp. 46-46. Emphasis in the original.
[6]Francisco, 2007. pp. 159-160.
[7]Rosenberg and Trevathan, 1995.
[8]Shipman, 2013, p. 426.

The too Rapid Evolution Claim Examined

In contrast to this common claim, Professor Bromhall writes, "it is simply not true that the [human] brain grows so fast that it 'forces' the baby to be born before it is ready," as is often claimed by evolutionists.[9] In fact, the "vast majority of the size difference between human brains and those of other primates results from a *far longer period of growth after the baby is born.*"[10] Chimp's brains nearly double in size between birth and adulthood whereas, in contrast, the human brain quadruples in size and does not stop growing until around age twenty.

Professor Bromhall concludes biological research has documented the human brain does *not* grow as fast compared to most animals. The fact is, "the human body develops at an incredibly slow rate—it takes far longer [for humans] to progress from one developmental stage to the next than any other primate."[11] So much for the Darwinists claims. Actually, humans as a whole have *fewer* birthing problems due to brain size and birth canal ratio issue than many primates. The many examples include squirrel monkeys who must actively help with their own birth.[12]

The Fossil Record

The large number of fossil finds of claimed human ancestors in the past 20 years has seriously questioned the *obstetrical dilemma* theory.[13] A problem is the "hominin fossil record, pelvic remains are sparse and are difficult to attribute taxonomically when they are not directly associated with crania."[14] The existing pelvis design is not due to childbirth constraints, and if "birth constraints do not explain pelvic variation of the past, that must mean we really do not understand why human pelvises look the

---

[9]Bromhall, 2003. p. 180.
[10]Bromhall, 2003. p. 180.
[11]Bromhall, 2003, pp. 180-181.
[12]Bromhall, 2003. Pp. 179-180.
[13]Churchill, and VanSickle. 2017.
[14]VanSickle, 2016a. pp. 321-322.

way they do today."[15]

One likely possibility is the differences in pelvis shapes found in the fossil record have nothing to do with evolution, but rather reflect age and sex differences. Research has found that, instead of evolution, "pelvis shape differs at different ages, even among adults, and to a greater extent in women than men."[16] Furthermore, health and nutrition variations need to be considered, a factor generally ignored, usually because it is very difficult to determine this concern from fossil bones. Paleoanthropologists have spent far too much time on attempting to explain morphology differences by evolution, ignoring other factors that can make a critical difference in morphology.[17]

VanSickle completed her PhD dissertation at the University of Michigan on this very topic. She reviewed the problems with *obstetrical dilemma*, noting there is more variation in pelvic dimensions than there is in neonatal head size.[18] She found that "Neandertal females more closely resemble Neandertal males than they do females today" indicating that the female birth canal in modern humans is comparatively larger in modern humans. She concluded that, "it may be that the observed pelvic differences, regardless of how they actually affect childbirth, may be caused by differences in locomotion."[19] Problems in her research include only seven female Neandertal individuals were identified, and these were in poor condition and/or incomplete. Another claim is that "hominins have been bipedal for so long that any obstetrical compromises required would have evolved long ago."[20]

The Female Pelvis Designed for Childbirth

The female pelvis, Latin meaning 'basin,' is a bone cradle that holds, and even allows, the baby to rock while it is still developing in the uterus. The

---

[15]VanSickle, 2016. p. 361.

[16]VanSickle, 2016. p. 361.

[17]VanSickle, 2016. pp.17- 21, 32-37, 40-42, 137-147, 361.

[18].VanSickle, 2014. p. 20.

[19]VanSickle, 2014. p. 121.

[20]VanSickle, 2014. p. 36.

Figure 5.1: The Female Pelvic Bones

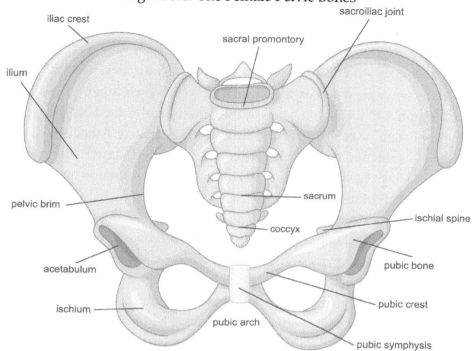

Image Credit: Rendix Alextian / Shutterstock.com

pelvis consists of two large hip bones that form the sides of the cradle and meet at the front of the body. They are connected by the symphysis pubis, a cartilaginous joint that, during pregnancy, a hormone called relaxin causes it to become flexible, allowing the birth canal opening to increase in size.[21]

During pregnancy, hormones also cause the ligaments between two other joints, the sacroiliac joints located between the sacrum and each iliac bone which forms the hips, to soften, causing a slight separation of the joints. This allows the required flexibility for the baby's head to pass through during birth. As one doctor wrote, the pelvis "is amazingly designed for its functions, especially for giving birth."[22] Some birthing problems are more a result of the mother's attitude about the birth process

---

[21] Isreal and Groeber 1960. p. 2-8.
[22] http://www.pregnancy-and-giving-birth.com/female-pelvis.html.

and her fears, not the physical design of the birthing system.[23]

In some cases, the pelvis is distorted from the optimal design called the Gynecoid Pelvis where the pelvic brim is close to a circle instead of an oval. Other problems, most of which are actually rare, include a very petite woman who gives birth to a very large child, or a very young or an older woman who has a distorted pelvis. When a baby is so large that it cannot fit through the pelvis, the condition is termed cephalopelvic disproportion. According to the American Pregnancy Association, true cases of cephalopelvic disproportion are very rare today.[24] The highest estimate I have seen is less than 1 out of 1,000 mothers have an unfavorable birth canal child head ratio.[25]

A detailed study of the problem including a meta-analysis and an analysis of new data, found many Cesarean sections are unnecessary and the procedure can increase the risk of a wide variety of health problems both for the mother and the child.[26] The report added that, the rate of Cesarean sections greatly varies across providers, facilities and states indicates this operation is often unnecessary. The report concluded, if the mother was aware of the risks of a Cesarean section, more women would opt for a normal delivery.

Cephalopelvic Disproportion Diagnosis

Many mothers whose labor process is labeled "failure to progress" are often incorrectly given a cephalopelvic disproportion diagnosis. However, when this problem can be confirmed as the delivery problem, the recommended safest delivery protocol is a Cesarean section.[27] The most common causes today of cephalopelvic disproportion are major congenital abnormalities or severe injuries, such as a pelvic fracture due to a sports or traffic accident. In the past, due to illnesses such as rickets and polio, pelvic anomalies causing birthing problems were far more common

---

[23]Beck, 2004. Pp. 28-35.
[24]https://www.bellybelly.com.au/birth/small-pelvis-big-baby-cpd/.
[25]Shipman, 2013, p. 426.
[26]Henic. 2012.
[27]Impey, and O'Herlihy. 1998.

than today.

The skull of the newborn is fairly flexible and, in most cases, able to conform to the birth canal opening. An infant's skull consists of six separate cranial bones held together by strong, fibrous, elastic sutures. The spaces between the bones that remain open in babies, the fontanelles, are a normal part of child development. The posterior fontanelle closes by age 1 or 2 months and the anterior fontanelle closes between 9 and 18 months. The flexibility of the sutures allows the bones to overlap so the baby's head can pass through the birth canal without pressing on, and damaging, the brain. During infancy and childhood, the sutures and fontanelles are flexible, allowing the child's brain the ability to develop properly and still protect the brain from minor impacts.

The problems outlined above have caused researchers to challenge the *obstetrical dilemma*. One reason is the finding that human gestation is shorter compared to primates, so birth occurs when the child's head is smaller. If the baby was in the womb longer, its head would grow even larger, creating more big head-small birth canal disproportion problems. A recent study found the duration of human pregnancy is 38-40 weeks compared to 32 weeks for chimps and 37-38 weeks for the great apes.[28]

Another reason for rejecting the *obstetrical dilemma* thesis is birth occurs when the fetus's growth reaches a certain size, thus the vast majority of births occur *before* cephalopelvic disproportion is a problem.[29] When the fetus reaches a certain size, the mother can no longer provide for the child's nutrient needs, a theory called the *energetics hypothesis*. The rare case in which cephalopelvic disproportion occurs is what needs to be explained.

Summary

Darwinists often imply that the female pelvis is too small in a large number of deliveries today, when in fact, this problem is very rare, and is mostly due to abnormal pelvic distortions. Furthermore, the female pelvis is

---

[28]Shipman, 2013, p. 427.
[29]Dunsworth, H. et al., 2012.

Figure 5.2: The Newborn Skull

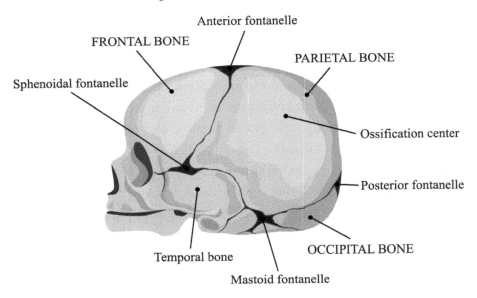

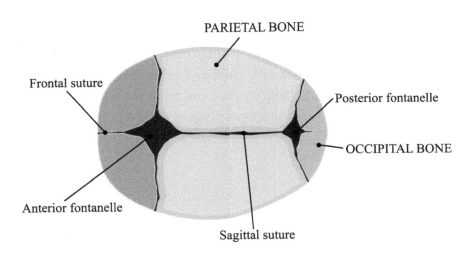

Image Credit: logika600 / Shutterstock.com

designed to effectively serve several very different functions, including stable support for the internal organs, and a design allowing for effective bipedal walking which requires the hips to be a certain shape, as well as delivery of the baby when he/she has developed sufficiently.

Its design must accommodate all of these very diverse, somewhat contradictory, requirements. Furthermore, the mutational load that has accumulated in the past 6,000 years has had an effect making comparisons today only imperfectly reflecting the original design. This reminds one of Genesis 3:16, where God said to the woman, as a result of sin, '"I will make your pains in childbearing severe; with painful labor you will give birth to children."[30] This is in contrast to animals. Professor Smith wrote:

> Having been raised on a farm, I have witnessed the birth of many animals, including, dogs, cats, cattle, horses, sheep, goats, pigs and others. Unless there are complications, the birthing process seems relatively pain free with the mother immediately cleaning the newborn and helping it nurse.[31]

---

[30]NIV version (the KJ is fine as well).
[31]Smith and Stockburger, 2018.

References

American College of Obstetricians and Gynecologists. 2016. *Your Pregnancy and Childbirth. Revised sixth edition.* Washington, D.C.

Angier, Natalie. 1999. *Women: An Intimate Geography.* Boston: Houghton Mifflin.

Ayala, Francisco. 2007. *Darwin's Gift to Science and Religion.* Washington, D. C. Joseph Henry Press.

Beck, Cheryl Tatano. 2004. Birth Trauma: In the Eye of the Beholder. *Nursing Research.* 53(1):28-35. January-February.

Bromhall, Clive. 2003. *The Eternal Child: Staying Young and the Secret of Human Origins and Behaviour.* 2003. Ebury Press.

Churchill, Steven and Caroline VanSickle. 2017. Pelvic Morphology in Homo Erectus and Early *Homo.* *The Anatomical Record.* 300:964-977.

Dunsworth, H. et al., 2012. Metabolic Hypotheses for Human Atriciality. *Precedings of the National Academy of Science of the USA* 109:15212-15216.

Dvorsky, George. 2014. The Most Unfortunate Design Flaws in the Human Body https://io9.gizmodo.com/1518242787.

Hafer, Abby. 2016. *The Not-So-Intelligent Designer: Why Evolution Explains the Human Body and Intelligent Design Does Not.* Cascade Books. Chapter 11 Bad Design—The Birth Canal. Pp. 45-52.

Henic Goer. 2012. Vaginal or Cesarean Birth: What Is at Stake for Women and Babies? *A Best Evidence Review.* http://www.nationmaster.com/country-info/stats/Health/Births-by-caesarean-section

Impey, Lawrence and Colm O'Herlihy. 1998. First Delivery After Cesarean Delivery for Strictly Defined Cephalopelvic Disproportion. *Obstetrics & Gynecology,* November.

Isreal, Leon and Groeber, Walter. 1960. Relaxin: A Persistent Challenge. *Obstetrics & Gynecology:* January 15(1):2-8.

Lents, Nathan. 2018. *Human Errors: A Panorama of our Glitches, From Pointless Bones to Broken Genes.* Boston, MA: Houghton Mifflin.

Moxon, Seve. 2003. Review of *The Eternal Child: An Explosive New Theory of Human Origins and Behaviour* by Clive Bromhall. *Human Nature Review* 3: 402-405.

Natalie Angier. 1999. *Women: An Intimate Geography*. Boston: Houghton Mifflin.

Nygaard, Ingrid. Ed. 2008. *Danforth's Obstetrics and Gynecology*. 10th ed Lippincott, Williams and Wilkins.

Preventing the First Cesarean Delivery: Summary of a Joint Eunice Kennedy Shriver National Institute of Child Health and Human Development, Society for Maternal-Fetal Medicine, and American College of Obstetricians and Gynecologists Workshop.

Roizen, Michael and Mehmet Oz. 2009. *You Having A Baby*. New York: Free Press.

Rosenberg K, and W. Trevathan. 1995. Bipedalism and human birth: The obstetrical dilemma revisited. *Evolutionary Anthropology* 4:161–168.

Rosenberg KR. 1992. The evolution of modern human childbirth. *Yearbook Physical Anthropology*. 35:89–124.

Rosenberg K and W. Trevathan. 2002. Birth, obstetrics and human evolution. BJOG 109:1199–1206.

Shipman, Pat. 2013. Why is Childbirth so Painful? *American Scientist*. 101(6):426-429.

Smith, E. Norbert. and Robert Stockburger. 2018. Childbirth Pain. Medical and Biblical perspectives regarding the pain of giving birth. Paper in Press.

Spong, Catherine Y.; Berghella, Vincenzo; Wenstrom, Katharine D.; More. *Obstetrics & Gynecology*. 120(5):1181-1193, November 2012.

Trevathan, Wenda. 1987. *Human Birth: An Evolutionary Perspective*. New York: De Gruyter.

Thibodeau, Gary and Kevin Patton. 2009. *Anatomy and Physiology*. St Louis, MO: Mosby.

VanSickle, C. 2014 *A New Examination of Childbirth-Related Pelvic Anatomy in Neandertal Females*. PhD Thesis. The University of Michigan.

VanSickle C, Cofran ZD, Garcia-Martinez D, Williams SA, Churchill SE, Berger LR, Hawks J. 2016. Primitive pelvic features in a new species of *Homo*. *American Journal of Physical Anthropologists* 159(S62): 321.

VanSickle, Caroline. 2016. An Updated Prehistory of the Human Pelvis. *American Scientist*. 104(5):361.

VanSickle, Caroline. 2016a. Primitive pelvic features in a new species of Homo. Presentation at the 85th Annual Meeting of the American Association of Physical Anthropologists.

Vidal-Cordasco, M. ; Mateos, A: Zorrilla-Revilla, G; Prado-Nóvoa O; Rodríguez, J.M; Energy c ost of walking in fossil Hominins. *American Journal of Physical Anthropology*. 2017 164(3):609-622. November.

Chapter 6

The Male Reproductive System

One of the latest "proofs" of human evolution given by evolutionists is the poor design of the male reproductive system, specifically the location of the testicles. Rowe lists the male reproductive system as number four in his list of the top ten examples of poor design in the human body.[1] The conclusion is "if testicles were designed," then why didn't God "protect them better. Couldn't the Designer have put them inside the body, or encased them in bone" like the brain which is surrounded by a hard skull?[2]

Instead of asking *why* the existing design exists, Oxford University Ph.D. Professor Hafer claims that this body structure is poorly designed. The accolades she earned for her book touting this view are in Appendix B. This approach is a science stopper because the "why" question motivates research into the *reasons* for the design. When this approach was applied to the human appendix, the tonsils, the backward retina, and the many other putative examples of poor design, in all cases very good reasons for the existing designs were found. The same is true of the male reproductive system.

---

[1]Rowe. 2016, p. 2.
[2]Hafer A., 2016. p. 5; Lainer.

Hafer came to her conclusion first without attempting to determine the *why* concern, explaining that, when looking for new approaches to refute Intelligent Design, she knew she "had a winner when . . . in the middle of an Anatomy and Physiology lecture." She concluded that the male reproduction system "is a great . . . argument against ID."[3] She also believed that she had a good "political-style argument" against ID, clearly the wrong reason for doing science.[4] Her main argument is that because male testicles are located in a sack outside of the body, they are prone to injury. She adds that many animals, including cold-blooded reptiles, they are located *inside* of the body where they are fully protected.

Reasons for the Existing Design

Male human testicles exist outside of the body in most mammals for several important reasons. One of the many reasons include this design can achieve effective scrotal temperature regulation for optimal spermatogenesis development. Another reason is to keep sperm relatively inactive until they enter the warm confines of the female reproductive system.[5] Even only a few degrees above the optimal temperature are detrimental to both sperm production, specifically in the later stages of spermatogenesis, and sperm maturation.[6]

A low ambient temperature is essential for normal spermatogenesis in humans and *most* mammals because the enzymes required for sperm production are denatured if their temperature is not finely regulated within very narrow limits. One study of mice found temperatures 37C or higher caused "a significant reduction in the percentage of motile sperm," producing, among other problems, an increase in the number of spermatozoa with plasma membrane damage.[7]

The main mammal exceptions include monotremes, mammals that lay eggs instead of giving birth to live young. Other animals with intra-

---

[3]Hafer, 2015; Haffer. 2016.
[4]Hafer, 2015 pp. 1-2.
[5]Wechalekar, et al., 2010.
[6]Setchell, 2006.
[7]Wechalekar *et al.*, 2016. pp 591-592.

abdominal-testes include a few placental mammals, such as insectivores (shrews, hedgehogs and moles), plus elephants and hippopotami .[8]

Testes externalization is found in mammals whose lifestyle involves aggressive jumping, leaping or galloping because this behavior would be expected to put great pressure on the testicles in large animals with this lifestyle. It could even expel their contents by creating concussive hydrostatic rises in peritoneal pressure.[9]

Temperature Concerns in Humans

Compared to core body temperature, the average temperature drop achieved in the external testicular design is ideal for sperm production. The temperature must be maintained at a temperature very close to 4C cooler than the normal body temperature of around 37°C.[10] An increase in temperature by as little as 2°C, adversely affects sperm formation. One result of this 2C increase in humans includes a significantly lower sperm count, and a significant increase in the number of abnormal sperm.[11]

If the testes were inside the body, the enzymes that sperm require to be healthy would be denatured in a matter of hours. New sperm would have to constantly be produced to allow year-round fertility, as is normal in humans. This issue would not be a concern for most all animals that are fertile only during a very short window each year.

Several complex mechanisms exist to ensure that the 4C lower temperature is maintained despite a body core of 37 C. When testicle temperature drops 4 below 37 C, or 33C, a complex feedback system causes the cremaster muscle surrounding the testicles to contract, which moves the testicles closer to the warm 37C body to compensate for the heat loss.[12] When their temperature rises *above* 33C, the cremaster muscle relaxes, allowing the testicles to move *away* from the body. This insures that the

---

[8]Sodera, 2009. pp. 109-110.

[9]Chance, 1996.

[10]Hutson, and Clarke, 2007. p. 65.

[11]Ahmad et al., 2012.

[12]Van Niekerk, 2012.

ideal male reproductive system temperature is maintained within a very narrow tolerance of the 33C ideal.

The 33C temperature is also maintained by increasing, or decreasing, the surface area of the tissue surrounding the testicles, the scrotum, allowing faster or slower dissipation of their heat. It achieves this by expanding as a wrinkled balloon expands when air is forced into it. Furthermore, scrotal skin is very thin, allowing the testes to easily lose heat into the surrounding environment. The air circulating around the scrotum sack also significantly helps to facilitate the cooling of the scrotal skin, in turn helps to cool the sperm development mechanism.

Furthermore, to maintain the proper temperature, the arteries carrying blood *into* the scrotum run alongside the veins that carry blood *away* from the scrotum. This sophisticated heat-exchange mechanism *lowers* the temperature of the blood traveling to the testicles.[13] The warm arterial blood coming from the abdomen loses heat to the cooler venous blood moving away from the testes. The result is that the blood is cooled slightly before even entering the scrotum. For these and other reasons, the existing system to maintain optimal spermatogenesis temperature control is an excellent design, well-known to engineers as *counter-current exchange*. This design is widespread in biological systems.

Year-Round Fertility

One major reason for the rigid temperature regulation is because humans are fertile throughout the entire year in contrast to most all animals with internal testicles, including all cold-blooded vertebrates and birds. Most animals need to be fertile only during the very short period of time during their mating season. Mating season often occurs when outdoor temperature allows maintenance of the proper temperature for spermatogenesis, such as is the case for reptiles.

Another reason for the rigid temperature regulation is that sperm have a very short lifespan and must be stored at *lower* than normal body temperature to allow them to remain dormant longer. Sperm are stored in the

---

[13]Wechalekar, et al., 2010, p. 591.

epididymis where they mature. The warmth of the female reproductive system then serves to help activate them. If the testicles were located inside of the male body, the sperm would be activated much sooner, and thus, given their short life span, usually measured in hours, large numbers of sperm would die before they could even enter the vagina. In harmony with this observation, "taxa with internal testes produce large volumes of low-quality sperm while taxa with scrotal testes produce smaller volumes of higher-quality sperm."[14]

Evidence for the temperature effect includes the fact that in humans semen volume remains fairly consistent year round, but semen *quality* is lower in the warm summer months compared to the cold winter months. Consequently, the total viable sperm count falls in the summer compared to winter, especially in the northern hemisphere during the hottest summer months of July and August.[15]

Testicular Protection Designs

Several designs reduce the likelihood of testicular injury, including the left testicle usually hanging lower than the right one. As a result, leg pressure causes one to slip past the other without pain or injury. Each testicle is housed in a strong fibrous outer covering called the tunica albuginea, and an effective lubrication system allows testicular slippage to occur without pain or problems. Injury is rare, and the main source of injury is in certain contact sports, which is why it is recommended that male sport participants always use protective equipment, such as a jockstrap or hard cup, while playing.

Cryptorchidism

Another reason for the existing placement of the human male reproductive system outside of the body is that the postpartum testicle is designed

---

[14]Freeman, 1990, p. 429.
[15]Chen et al., 2003.

to function at the lower temperature achieved by the existing testicular-scrotum design. Failure of the testicles to descend into the scrotum, called cryptorchidism, causes a significant increased risk of malignancy and other major health problems. The complex process of testicle descent is also both complex and poorly understood."[16] Failure of the testicles to descend following birth leads to progressive abnormality of biochemistry and physiology of both testes, often at the least causing infertility.[17]

One example is that the abnormal biochemistry caused by descent failure interferes with many of the necessary reproductive system developments. Examples include the transformation of neonatal gonocytes into type A spermatogenesis, a step required to produce viable sperm.[18] This is one reason why failure of the testicles to descend a major reason for male infertility.

Evolution of the Scrotum Unknown

Evolutionists are forced to speculate on how the many complex male temperature regulative system parts could have evolved due to a complete lack of evidence. Their theories are based on neither observation nor empirical science, but rather arm chair speculation.[19] Consequently, evolutionists must produce just-so stories in an attempt to explain their origin and function.[20] In short, a literature review found that "all of the current hypotheses regarding the origin and evolution of the scrotum" and external testicles are seriously problematic.

Evolutionists assume external testicle evolution from lower life forms with internal reproductive organs. This assumption is problematic because it is "why the scrotum has been *lost* in so many groups, that should be explained."[21] Some evolutionists even speculate that the scrotum may have evolved *before* mammals, opining that

---

[16]Hutson and Clarke, 2007, p. 65.
[17]Chung and Brock. 2011.
[18]Hutson and Clarke, 2007, p. 66.
[19]Gallup, et at., 2009; Freeman,1990.
[20]Heyns, C. F. and Hutson, J. M., 1995.; Ivell, 2007.
[21]Werdelin, and Nilsonne, 1999. pp. 61-62.

in concert with the evolution of endothermy in the mammalian lineage, and that the scrotum has been lost in many groups because descent in many respects is a costly process that will be lost in mammal lineages as soon as an alternative solution to the problem of the temperature sensitivity of spermatogenesis is available.[22]

A review of the claim that the male reproductive system is poorly designed concluded that several major reasons exist for the testicular system as designed in humans, including the fact that it is required for viable sperm. Furthermore, the improvements claimed by Darwinists would not improve it, but rather would cause major fertility and other problems. The evolutionists cited in this chapter are not making scientific or logical arguments for poor design. They are instead advancing an "argument from personal incredulity." It is not demonstrated that the structures discussed above actually function poorly, only that the critic is attempting to object to the existing design based on the preconceived belief that they are poorly designed.

Conclusions

In conclusion, clear evidence exists that year-round reproductive cycles, plus the requirement that human sperm must be kept close to the constant temperature of 4C below that of the core body temperature, effectively explains existing testicular design. Men who have uncorrected non-descended testicles are usually infertile and prone to other health problems, including cancer. In short, the existing complex design is required for many reasons, including fertility and health reasons. It is, therefore, clear that Hafer's poor design claims, and those of other evolutionists, are grossly irresponsible.

---

[22]Werdelin and Nilsonne, 1999, p. 61.

References

Ahmad, Gulfam; Nathalie Moinard, Camille Esquerre -Lamare, Roger Mieusset, and Louis Bujan. 2012. Mild induced testicular and epididymal hyperthermia alters sperm chromatin integrity in men. *Fertility and Sterility*. 97(3):546-553. March.

Chance, M. R. A. 1996. Reason for externalization of the testis of mammals. *Journal of Zoology*. 239(4):691-695. August.

Chen Z., Toth T, Godfrey-Bailey L, Mercedat N, Schiff I, Hauser R. 2003. Seasonal variation and age-related changes in human semen parameters. *Journal of Andrology*. 24(2):226-31. March-April.

Chung, Eric and Gerald Brock. 2011 "Cryptorchidism and its impact on male fertility: a state of art review of current literature. *Canadian Urological Association Journal*. 5(3)210-214.

Freeman, Scott. 1990. "The Evolution of the Scrotum: A New Hypothesis." *Journal of Theoretical Biology*, 145:429-445.

Gallup, Gordon G., Mary M. Finn and Becky Sammis. 2009."On the Origin of Descended Scrotal Testicles: The Activation Hypothesis." *Evolutionary Psychology* 7(4):517-526.

Hafer, Abby. 2015. "No Data Required: Why Intelligent Design is Not Science." *The American Biology Teacher*, 77(7):507-513.

_____. 2016. *The Not-So-Intelligent Designer: Why Evolution Explains the Human Body and Intelligent Design Does Not*. Cascade Books, an Imprint of Wipf and Stock Publishers.

Heyns, C. F. and J. M. Hutson. 1995. "Historical Review of Theories on Testicular Descent." *Journal of Urology*. 153:754-767.

Hutson, John M. and Melanie C. C. Clarke. 2007. "Current Management of the Undescended Testicle." *Seminars in Pediatric Surgery*, 16:64-70.

Ivell, Richard. 2007. Lifestyle impact and the biology of the human scrotum. *Reproductive Biology and Endocrinology*. 5(15):1-8.

Lanier, Jaron. 2006. Jaron's World. *Discover*. 27(12):22-24.

Rowe, Chip. 2016. The 10 Design Flaws in the Human Body. *Nautilus*. March, 24.

Setchell, B. P. 2006. The effects of Heat on the Testes of Mammals. *Animal Reproduction*. 3(2):81-91.

Sodera, Vij. 2009. *One Small Speck to Man – The Evolution Myth. 2nd edition*. London: Vij Sodera Productions

Van Niekerk, E., 2012. Vas deferens—refuting 'bad design' arguments, *Journal of Creation*. **26**(3):60–67, 2012; http://creation.com/vas-deferens.

Wechalekar, Harsha, Brian P. Setchell, Eleanor J. Peirce, Mario Ricci, Chris Leigh, William G. Breed. 2010. Whole-body heat exposure induces membrane changes in spermatozoa from the cauda epididymis of laboratory mice. *Asian Journal of Andrology*. 12(4): 591–598.

Werdelin, Lars and Asa Nilsonne. 1999. "The Evolution of the Scrotum and Testicular Descent in Mammals." *Journal of Theoretical Biology* 196(1):61-72, January 7.

Chapter 7

The Prostate Gland

The common claim that the prostate is poorly designed was reviewed, concluding that there are many very good reasons for its existing design and also its location around the urethra. The main prostate problem, benign prostatic hypertrophy, is caused by disease, hormone imbalance, and other health problems, not poor design.

Enlargement of the prostate, a condition called *benign prostatic hypertrophy* (BPH) **or *benign prostatic hyperplasia*,** occurs to some degree in approximately half of all men by age 60. The most common result of BPH is blockage of urine flow in males. This problem is widely touted by Darwinists as proof of poor design and evidence that humans were not designed but rather evolved by natural selection of random mutations.

The Prostate Gland

The prostate in an average adult male is a walnut-shaped gland four cm long by two cm wide and weighing about 20 grams. Unique to males, it is small at birth, enlarges rapidly at puberty, and sometimes shrinks in octogenarians. Located directly below the bladder, it surrounds the urethra where the urethra exits the bladder (see Figure 7.1). The prostate gland itself is a complex fibromuscular structure that contains a conglomerate of 30 to 50 small saccular glands. The entire prostate is

Figure 7.1: Illustration of the Male Prostate

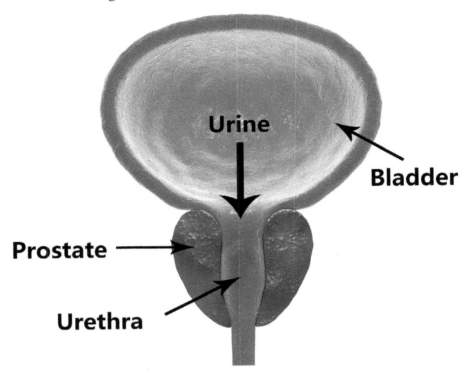

Image Credit: Kateryna Kon / Shutterstock.com

divided into two main lobes that are totally enclosed in a tough fibrous capsule.[1] The proximal section connected to the bladder is called the base, the opposite end, a narrow distal structure, the apex. The main divisions of the prostate are the central region, the transitional (inner) region which surrounds the urethra, and the peripheral, or outer layer.[2]

Among the prostate's several functions include producing the secretions that constitute much of the volume of the seminal fluid.[3] The secretions consist of a thin, opalescent liquid that contains a soup of complex compounds, including citric acid, phosphatase, prostaglandin (named after the prostate gland because it was first discovered there), diastase,

---

[1]Garnick, 1996, p. 27.
[2]Naz, 1997.
[3]Swanson and Forrest, 1984.

fructose, beta glucuronidase, a potent fibrinolysin, spermine, and several proteolytic enzymes.[4] Semen contains mucus which has an alkaline pH of 7 to 8 and serves to neutralize the acidic male urethra and female reproductive tract to allow sperm to survive their trip from the testicles to the uterine (fallopian) tubes.

The Poor Design Argument

A common dysteleological claim is that the human prostate gland is poorly designed because it surrounds the urethra, a thin tube resembling a miniature straw that carries urine from the bladder to exit the body.[5] An example of this "poor design" argument is the following claim:

> In human males, the urethra passes right through the prostate gland, a gland very prone to infection and subsequent enlargement. This blocks the urethra and is a very common medical problem in males. Putting a collapsible tube through an organ that is very likely to expand and block flow in this tube is not good design. Any moron with half a brain (or less) could design male "plumbing" better.[6]

University of Chicago Biologist Jerry Coyne wrote

> why did God—sorry, the Intelligent Designer—give whales a vestigial pelvis, and the flightless kiwi bird tiny, nonfunctional wings? ... What a joker! And the Designer doesn't seem all that intelligent, either. He must have been asleep at the wheel when he designed our appendix, back, and prostate gland.[7]

Professor Karen Bartelt wrote that, if we assume, as Intelligent Design postulates, that

---

[4]Bloom and Fawcett, 1969.
[5]Garnick, 1996, p. 26.
[6]Colby, et al., 1993, p. 1.
[7]Coyne, 2006, p. 1.

humans can discern design, then I submit that they can also discern poor design (we sue companies for this all the time!). In *Darwin's Black Box*, Behe refers to design as the "purposeful arrangement of parts." What about when the "parts" aren't purposeful, by any standard engineering criteria? When confronted with the "All-Thumbs Designer"—whoever designed the spine, the birth canal, the prostate gland, the back of the throat, etc. Behe and the ID people retreat into theology.[8]

The on-line encyclopedia Wikipedia (2007), under the heading "Argument from poor design, a dysteleological argument against the existence of God," listed as the first example of poor design the "urinary tract in the human male, especially the unnecessary passage of the urethra through the prostate gland. As the prostate almost always grows with age, it eventually compresses the urethra and often makes urination difficult or even impossible." Contrary to anti-Intelligent Design proponents, this claim assumes that you *can* detect design in biological systems.

Evaluation of the Poor Design Claim

It is true that, when the prostate becomes diseased or enlarges as males approach retirement age, it can interfere with urine flow. Also, when the prostate enlarges, it obstructs the urethra, distorts the bladder neck outflow tract, and interferes with the normal sphincter mechanisms so that bladder evacuation can become incomplete. This condition causes the patient to assume that he has emptied himself, but some urine always remains in the bladder. This results in the problem that requires needing to urinate again after only a short time after passing urine.

Urine flow blockage in males is usually caused by prostate enlargement, a medical condition termed *benign prostatic hypertrophy* (BPH). Although some enlargement occurs in about half of all men by age 60, the condition often can be successfully treated with medication, and in

---

[8]Bartelt, 1999, p. 4.

about 70 percent of all cases the enlargement is minor, and the number of moderate to severe cases increases with age.[9]

When enlargement interferes with normal everyday activities, this condition is called clinical BPH. One study found only 25 to 30 percent of men in the 70 to 74 age range had clinical BPH.[10] The prostate is not "likely to expand," nor does it enlarge because of poor design, but rather it usually enlarges as a result of poor health, bacterial or viral infection, or other diseases, genetic mutations, hormonal imbalance, race or ethnicity influences, poor diet, smoking, sexual habits, obesity, the excess use of certain medications, abnormal androgen production levels, lack of testosterone regulation, or cancer.[11]

Nor is blockage the result of poor design, but rather is a clear early indication that something is wrong—such as prostate cancer—and that the affected person should consult a physician to determine the cause. It is for this reason that changes in urine flow often trigger an examination that allows early detection of prostate cancer, usually by a Prostate Specific Antigen (PSA) evaluation and by digital examination of the prostate, and thus yields a higher level of probability of successful treatment.[12] Cancer cells often originate in the posterior region of the prostate, hence early cancer can be felt as very hard nodules on rectal examination.

Causes of Benign Prostate Hypertrophy

BPH occurs mainly in older men and does not develop in men who were castrated before puberty. For this reason, factors related to aging and the misregulation of hormones produced by the testes are believed to be major causes of BPH. Men produce both testosterone and small amounts of the female hormone estrogen throughout their lives. As men age, the amount of active testosterone in the blood decreases, leaving a higher proportion of estrogen.[13] Animal studies suggest that BPH can

---

[9]Naz, 1997, p. 9.

[10]Naz, 1997, p. 10.

[11]Naz, 1997.

[12]Foster and Bostwick, 1998.

[13]Snyder, et al., 1999.

result from an abnormal amount of estrogen within the gland and this condition increases the activity of cell growth promotion hormones.

Another cause focuses on a testosterone derived substance called dihydrotestosterone (DHT) that may help to regulate prostate growth. Most animals lose their ability to produce DHT as they age. Even with a decrease in blood testosterone levels, some older men continue to produce abnormally high levels of prostate DHT. High levels of DHT and other hormones such as Oxytocin may encourage abnormal cell division, causing BPH.[14] Evidence for this conclusion includes the finding that men who produce normal DHT levels rarely develop BPH.

Research also suggests that BPH can develop due to "instructions" given to cells early in life. This theory concludes that BPH occurs because cells in one section of the gland "reawaken" later in life as a result of these instructions. These "reawakened" cells then deliver signals to other cells in the gland, either instructing them to grow or making them more sensitive to certain growth hormones.[15]

Reasons for the Existing Urethra Designs

One reason for the existing design is the prostate serves both as a support for the bladder and a spacer between the bladder and the urogenital diaphragm. This allows the vas deferens and the seminal vesicles room to connect to the urethra. Because the urethra is a small narrow tube without a substantial support system, the bladder, when full, would cause the urethra to kink. One possible solution could be bladder support by ligaments to prevent its interfering with possible urethra outflow, but this requires both a new support system, new attachment sites, and other structures. Economy allows the prostate to serve this role as well as the other roles noted in this chapter.

Rather than the urethra going through the prostate, it is more accurate to describe the prostate as an expanded part of the urethral wall. Each of the distinct glandular regions of the prostate develops from a different

---

[14]Cook and Sheridan, 2000; Nicholson, 1996.
[15]Nicholson, 1996.

segment of the prostatic urethra.[16] The fact that the prostate is part of the urethra is a good example of Intelligent Design because this design allows the prostate to more rapidly deposit its secretions into the urethra where they mix with the seminal vesicle secretions.

The prostate's 30 to 50 glands empty into 16 to 32 small excretory ducts that open independently into the urethra to enable it to rapidly transfer their products.[17] This rapid transfer is required in order for the system to operate effectively. The prostate's smooth muscle is a highly effective system to efficiently empty the tubules during the sexual response, thus forcing semen rapidly into the urethra. The tubules open on both the right and left sides of the colliculus seminalis, a U-shaped structure that runs along the urethra.

The entire prostate contracts during the sexual response, in turn contracting the U-shaped urethra, thus forcing semen forward in the urethra. Because the many small cavernous holes all open directly into the U-shaped opening, prostate contraction helps to effectively propel the semen forward during the sexual response. It must travel out of the urethra with sufficient force to make its journey successfully in order to achieve fertilization. In the words of Baggish, the urethral segment "of the capsule squeezes fluid out" of the prostate during ejaculation.[18] The fact that the normal healthy prostate is composed of 40 percent smooth muscle tissue, 40 percent connective tissue, and 20 percent glandular tissue, demonstrates the importance of this muscular function.

The rhythmic contractions of the prostate propel the semen forward in the urethra during the first stage of the sexual response.[19] A one-way valve called the preprostatic sphincter, prevents retrograde flow of semen. This "complex and intricate" structure is very "similar in different mammalian species and is highly conserved," which is evidence that the system is well-designed, highly effective, and fully functional.[20]

Those claiming that the prostate is poorly designed must have a better

---

[16]McNeal, 1998, p. 19.

[17]Foster and Bostwick, 1998, p. 3.

[18]Baggish, 1996, p. 2.

[19]Katchadourian, 1989, p. 71; deGroat and Booth, 1980; and Dunsmuir and Emberton, 1997, p. 319.

[20]Foster and Bostwick, 1998, p. 3.

alternate design in mind. The only other option is a separate prostate located to one side of the urethra. This would require a new duct system design that would add a cumbersome set of structures and greatly slow the flow of semen into the urethra. This alternative design would require another system to propel the semen into the urethra and yet another system to help propel the semen along the urethra to replace the several roles that the prostate now serves.

The Nerve Plexuses

The prostate also contains abundant complex nerve plexuses, mostly consisting of non-myelinated nerve fibers and sensory nerve endings. As noted, the prostate itself contracts during sexual excitement, and it is this contraction that produces much of the pleasure accompanying sexual activity due to the nerve plexuses. Autonomic nerves arising from the hypogastric plexus contribute to the prostate plexus. The normal healthy prostate is, for this reason, a major source of pleasure rather than of pain.[21]

Other Prostate Functions

Another important function of the prostate is to help control urine flow. A Cleveland clinic urologist report concluded that the "prostate gland, which surrounds the tube that allows urine to flow outside the body, also helps to hold back urine until the time to release."[22] When the prostate is removed, most men experience some degree of incontinence, usually lasting from three to six months, but in up to ten percent of cases lasting for the rest of their lives.[23] Post-prostatectomy incontinence levels can range from "total" to incontinence occurring only under certain conditions, such as in the stress incontinence condition. About ten percent of all men experience total urinary incontinence after prostate surgery. This is often

---

[21]Morganstern and Abrahams, 1996, p. 4.
[22]Shuman, 2006.
[23]Dierich and Felecia, 2000, p. 86.

temporary, but may last for years.[24] As Leach notes, "loss of bladder control after prostate surgery is a devastating complication, which has a significant negative impact on quality of life."[25]

The major cause of post-operation prostatectomy incontinence is disruption of the involuntary sphincter muscle at the bladder neck. This injury from the surgery can be permanent or temporary due to swelling, leaving only the voluntary lower sphincter muscle. Lack of the prostate after surgery, though, also contributes to the problem.

Harvard Medical School trained physician Andrew Weil wrote concerning prostatectomy that:

> Surgical removal of the prostate gland is another expensive and painful operation, frequently resulting in impotence and urinary dysfunction. It is done as treatment of benign prostate hypertrophy (BPH) and early stages of prostate cancer. In the case of cancer, removal of the gland is often unnecessary if the cancer is not aggressive.[26]

Because females lack a prostate to help control urine flow, some researchers conclude this is one of several reasons why women are much more likely to suffer from incontinence. Most incontinence in women results mainly from sagging (prolapse) of the anterior vaginal wall initiated by the trauma and stretching due to childbirth labor.

Are Any Prostate Accessory Structures Vestigial?

As scientists study the prostate, they have learned that it is a complex, well-designed structure with many accessory structures, all of which are important, and none of which are vestigial as once taught by Darwinists. This brief review can outline only a few of the major structures and functions of the prostate itself. An example of one structure is the *utriculus*

---

[24]Newman, 1997, pp. 85-86; Burgio, et al., 1990, p. 145.
[25]Leach, 2004, p. 1.
[26]Wallenchinsky and Wallace, 1993, p. 117.

prostaticus. Once thought to be useless, it is now known to be an important accessory gland.[27]

Summary

Evolutionists claiming that the prostate is poorly designed usually have only a very shallow knowledge of its many functions. To prove the poor design claim, evolutionists must come up with a better design that deals equally well with its many functions and requirements, a very unlikely scenario. From the existing evidence we must conclude that the prostate is well-designed for its many roles. The problem is not poor design, but infections, mutations, teratogens, degeneration of the genome, and unhealthy behavior.

Board specialist in pathology, Agatha Thrash, wrote that for all the prostate's "lowly credentials," it is "an amazing piece of engineering."[28] The prostate's several functions include producing the conditions (including the required pH) allowing sperm to survive their trip from the testicles to the fallopian tubes for fertilization. It serves all of these tasks remarkably well. After the child bearing years are past, the prostate can cause problems, but from a clinical standpoint, it is remarkably free of disease even during these later years. Only in the fifth or sixth decade of life does it cause problems, usually when long past its primarily reproductive function.

---

[27]Bloom and Fawcett, 1969, pp. 718-719.
[28]Thrash, 2006, p. 1.

References

Baggish, Jeff. 1996. *Making the Prostate Therapy Decision*. Los Angeles, CA: Lowell House.

Bloom, William and Don Fawcett. 1969. *A Textbook of Histology*. Philadelphia: Saunders.

Bartelt, Karen. 1999. "A Central Illinois Scientist Responds to the Black Box." *The Real News*, 7(12):1,4, December.

Burgio, Kathryn L., K. Lynette Pearce, and Angelo J. Lucco. 1990. *Staying Dry: A Practical Guide to Bladder Control*. Baltimore, MD: The Johns Hopkins University Press.

Colby, Chris and Loren Petrich. 1993. "Evidence for Jury-Rigged Design in Nature." *The Talk.Origins Archive*. `http://www.talkorigins.org/faqs/jury-rigged.html`.

Cook, Terry and William P. Sheridan. 2000. "Development of GnRH Antagonists for Prostate Cancer: New Approaches to Treatment." *The Oncologist*, 5:162-168.

Coyne, Jerry. 2006. Ann Coulter and Charles Darwin. Coultergeist. *The New Republic Online*. July 31, 2006.

deGroat, W. C. and A. M. Booth. 1980. "Physiology of Male Sexual Function." *Annuals of Internal Medicine*. 92(2):329-331.

Dierich, Mary and Froe Felecia. 2000. *Overcoming Incontinence*. New York: Wiley.

Dunsmuir, William and Mark Emberton. 1997. Surgery, drugs, and the male organism. editorial in the *British Medical Journal*. 314:319.

Foster, Christopher S. and David G. Bostwick. 1998. *Pathology of the Prostate*. Philadelphia, PA: Saunders.

Garnick, Marc B. 1996. *The Patient's Guide to Prostate Cancer: An Expert's Successful Treatment Strategies and Options*. New York: Penguin Books.

Katchadourian, Herant A. 1989. *Fundamentals of Human Sexuality*. New York: Holt, Rinehart, and Winston. 5th edition.

Leach, Gary. 2004. "Incontinence Treatment Options for Post-Prostatectomy. *Prostate Cancer Research Institute Newsletter*.

7(2)1-3. May.

McNeal, John. 1998. "Anatomy and Normal Histology of the Human Prostate." Chapter 2, pp. 19-34 in Foster and Bostwick.

Morganstern, Steven and Allen Abrahams. 1996. *The Prostate Sourcebook.* Updated Edition. Los Angeles, CA: Lowell House.

Naz, Rajesh K. (editor). 1997. *Prostate: Basic and Clinical Aspects.* Boca Raton, FL: CRC Press.

Newman, Diane Kaschak with Mary K. Dzurinko. 1997. *The Urinary Incontinence Sourcebook.* Los Angeles, CA: Lowell House. Foreword by Ananias C. Diokno.

Nicholson, Helen D. 1996. "Oxytocin: a Paracrine Regulator of Prostatic Function." *Reviews of Reproduction,* 1:69-72.

Shuman, Tracy. 2006. *Prostate Cancer: Urinary Incontinence.* The Cleveland Clinic Report, Cleveland, OH.

Snyder, P.J., H. Peachey, P. Hannoush, J. A. Berlin, L. Loh, D.A, Lenrow, J.H. Holmes, A. Dlewati, J. Santanna, and C.J. Rosen. 1999. "Effect of Testerone Treatment on Body Composition and Muscle Strength in Men Over 65." *Journal of Clinical Endocrinology and Metabolism,* 84(8):2647-2653.

Swanson, Janice M. and Katherine A. Forrest (editors). 1984. *Men's Reproductive Health.* New York: Springer Publishing Company. Springer Series: Focus on Men, Volume 3.

Thrash, Agatha. 2006. "Prostate Disease Counseling Sheet." Seale, AL: Uchee Pines Institute.

Wallenchinsky, David and Amy Wallace. 1993. *The Book of Lists: The '90s Edition.* Boston, MA: Little, Brown and Company.

Wikipedia. 2006. "Argument from poor design." https://en.wikipedia.org/wiki/Argument_from_poor_design. Accessed March 1, 2018.

Chapter 8

The Vas Deferens

The *vas deferens*, Latin for 'carrying-away vessel' (plural: *vasa deferentia*), often also called the *ductus deferens*, is a tube in the male reproductive system that carries sperm from the testes to the urethra for insemination.[1] The poor design claim is that the existing design is deficient because it does not take the shortest route from the testes to the penis, but instead loops around the bladder. A typical example is Richard Dawkins' claim that the vas deferens, "the pipe that carries sperm from the testes to the penis, . . . takes a ridiculous detour around the ureter, the pipe that carries urine from the kidney to the bladder."[2]

The late biologist George C. Williams, admits that he is "not aware of its [current design] causing medical problems," but claimed that the vas deferens was an example of the "functional absurdity in the male reproductive system."[3] Williams speculated that its poor design resulted because the testicles have moved downward during the course of evolution "from deep inside the body to the scrotal sac behind the penis," and that as "the testicles moved ever closer to the point at which they drain into the urethra, it might reasonably be expected that ever shorter tubes would be needed to convey the semen to its destination."[4] He opined that

---

[1]Knobil ,et al., 1993.
[2]Dawkins, 2009, p. 364.
[3]1997, pp. 138-140.
[4]Williams, 1997, pp. 138-139.

evolution is "concerned only with what is slightly more adaptive now, and is utterly blind to future consequences of current changes." Williams uses the following analogy to explain what occurred:

> the evolution of the mammalian reproductive system can be seen in the behavior of the gardener [see figure 2]. . . . He has watered his perennial border to the right and along the back fence, and now wants to continue to the left. Unfortunately the garden hose has caught on the tree. All he has to do is carry the nozzle clockwise around the tree, come back to his present position, and continue watering. But suppose he seeks an additional segment of hose to add to the length he already has. Stupid? Indeed.[5]

Williams concluded that "this is precisely the mistake that evolution made in moving the testicles from a central abdominal position to the scrotal position" because Williams claims the vas deferens tube gets hung up on the ureter, consequently it requires an excessively longer route than necessary.[6] Neither Dawkins nor Williams explains why 'evolution' supposedly achieved this poor design, but Dawkins, in his quote below, rather uncritically repeated William's claim that

> the vas deferens unfortunately got hooked the wrong way over the ureter. Rather than reroute the pipe, as any sensible engineer would have done, evolution simply kept on lengthening it—once again, the marginal cost of each slight increase in length of detour would have been small. Yet again, it is a beautiful example of an initial mistake compensated for in a post hoc fashion, rather than being properly corrected back on the drawing board.[7]

Another example of the same claim from a anti-Intelligent Design web site is as follows:

---

[5]Williams, 1997, pp. 138-139.
[6]Williams, 1997, p. 139.
[7]Dawkins, 2009, p. 365.

> Another instance of 'bad design' would be the path of the vas deferens in humans ... the vas deferens loops over the ureter and round in a rather long winded fashion. It doesn't take a great deal of intelligence to see the most optimal path for the vas deferens to take (straight down to the testicles). Why would a designer incorporate something so pointless and unnecessary into its design?[8]

The anonymous author concluded that only evolution can explain this bad design, namely humans "evolved from creatures whose testicles were in a different position, and when they 'moved' to their current position they looped over the ureter, rather than passing underneath it. [This is yet] another (rather embarrassing) problem for our designer."[9]

These claims totally fail to take into consideration both how the male reproductive system develops and functions. Part of the problem is "The vas deferens (ductus deferens) is the least understood and studied organ of the male reproductive system."[10] It was only in 1972 that one of the first detailed modern reviews of the epididymis was published and its related other structures.[11] For example, the complexity of the vas deferens is indicated by the discovery that the innervation of the vas deference involves the adrenergic, purinergic and peptidergic systems.[12] Conversely, as reviewed below, anatomists have documented that several critical reasons exist for the existing vas deferens' design.

The Vas Deferens' Design

The vas deferens is a vital part of male mammalian anatomy. Men are often sterilized by cutting this tube and sealing the cut ends, a procedure called a vasectomy. A fundamental implication of inept design is that a poorly engineered system by definition does not function very well,

---

[8]Anonymous, 2011.
[9]Anonymous, 2011.
[10]Lychkova and Puzikov, 2015; 2:1475.
[11]Robaire and Hermo, 1988, p. 99.
[12]Robaire and Hermo, 1988, p. 1004.

Figure 8.1: The Vas Deferens and its Extended Route to the Urethra

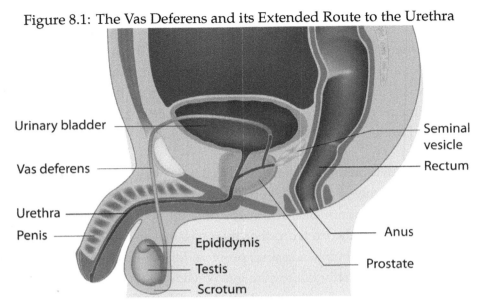

Image Credit: Alila Medical Media / Shutterstock.com

or not at all, due to the deficient design. Dawkins naively asserts that an engineer would not design a vas deferens with such a long detour and concludes that argument alone is evidence of poor design. This is a fallacious argument from personal incredulity.

Those who make the poor design argument fail to demonstrate how the design actually disadvantages the male. The fact is, the system works very well; and if it performed poorly due to its design, then natural selection would favor the putative superior design that Dawkins proposed. As will be documented, the direct route Dawkins' proposes would create several major problems.

Embryological Requirements

Many adult features are a result of embryological and fetal developmental requirements.

Likewise, the existing vas deferens design is partly a requirement of fetal development. Multi-cellular mammals begin as a fertilized egg (the zygote), which becomes an embryo, and next, a fetus. To produce an adult organism, the needs of the growing embryo must be satisfied at every developmental stage.

In view of this fact, the vas deferens design is actually ingenious: as the fetus elongates during development, moving the vas deferens over and around the ureter allows the proper length to be achieved when development is completed.[13]

The testes and ovaries also both develop from the same structure, the urogenital ridge, which is located proximal to where the kidneys begin to develop. Later, factors including hormone production, drive cellular differentiation. The male sex-determining gene on the Y chromosome controls testosterone and dihydrotestosterone secretion levels, and anti-Müllerian hormones cause suppression of Müllerian duct system development that later degenerates in males.[14]

The male testes are joined via the gubernaculum's testis cord.[15] The **gubernaculum testis set** arises after the mesonephrons have atrophied during the 7th week. Between the 7th and the 12th week, the **gubernaculum shortens,** pulling the testes, the vas deferens, and its vessels downward. As the fetus develops, the scrotum gubernaculum pulls the testes farther away from its origin until they reach into the scrotum roughly about the time of birth, and eventually penetrate through the body cavity. They are held in place by the **suspensory** ligament, the gubernaculum, and the scrotum. The vas deferens also follows this same route.[16]

The Vas Deferens' Design for Temperature Regulation

One important reason for the vas deferens' much longer course is due to the fact that sperm are very temperature sensitive.[17] Healthy male human gametogenesis (sperm formation) requires a narrow temperature range about four to seven degrees below the abdominal core temperature.

---

[13]Blechschmidt and Freeman, 2004.
[14]Carlson, 1996, p. 597.
[15]Johnson, 2007, p. 13.
[16]Romer, and Parsons. 1986.
[17]Setchell, 2006; Wechalekar, et al., 2010, p. 591; Lychkova and Puzikov, 2015.

[18] Undescended testes, a medical abnormality called cryptorchidism, due to body core heat damage, causes abnormal testicular metabolism, producing infertility and possibly tumors in the undescended testes.[19]

To regulate temperature, the testes move closer to the body to raise the gametogenesis temperature, and farther away from the body to lower the gametogenesis temperature. This control is partly due to the scrotum cremaster muscle that surrounds the testis, and also the *tunica dartos* consisting mainly of the **dartos muscle** located in the interior of the scrotum. Higher temperature causes the cremaster muscle system to relax, allowing the testes to move farther away from the body core to ensure that their temperature is about four to seven degrees below body core temperature. Cooler testicular temperatures cause the muscle to contract, drawing the testes in closer to the body. The testes also move closer to the body during sexual arousal to help protect them. The vas deferens, which is attached to the epididymis in the testes, must be long enough to accommodate this movement.

Another division of the testes temperature regulation system is a network of many small veins called the pampiniform plexus that surrounds the vas deferens. This ingenious heat exchange system is designed to cool the blood that supplies the testes to ensure that the testes are maintained at the required number of degrees below core body temperature. If the vas deferens followed a direct route to the penis, as Dawkins argues that "any sensible engineer would design," this vital temperature regulatory system would not be possible.

The existing design results in a vas deferens about 45 cm (18 inches) long, which allows the flexibility required to permit the testes to move closer, or farther away, from the body.[20] This is significant because temperature regulation requires the testes to be able to move up and down by as much as two inches. Its length, as well as looping the 45-cm long vas deferens over the ureter, allows for this required level of flexibility.

Although spermatogenesis in most all mammals is inhibited by higher temperatures, much variety in how temperature lowering occurs exists

---

[18]Johnson, 2007, p. 14.
[19]Chung and Brock, 2011.
[20]Jones and Lopez, 2006, p. 113.

among mammalian species.[21] The only reason the testes must descend outside of the body in humans could not have been temperature only. Some mammals effectively reproduce with internal testes, and for a few such mammals, the higher temperature is less a concern. For example, the testes of egg-laying mammals called monotremes, including platypus and echidna, are *not* located in a scrotal sac outside of the body.

Furthermore, the testes of a number of higher placental mammals do not descend into a scrotal sac. Examples include both small mammals, such as shrews, hedgehogs, moles, and a few very large mammals, such as hyraxes, hippopotamuses and elephants.[22] The testes in armadillos, whales, and dolphins migrate only part of the way to the lower abdomen. In hedgehogs, moles, and some seals, the testes lodge in the inguinal canal and "in most rodents and wild ungulates, they retain mobility in the adult, migrating in and out of the scrotum."[23] As far as is known, the routing of the vas deferens in humans is consistent across all these other mammal species, including the higher primates and giraffes. This supports the functional reasons given in this chapter for the existing routing of the vas deferens.

Evolution would predict that the testes in primitive mammals would not have migrated, and in more advanced animals they would migrate part way, and in the most advanced mammals, such as primates, they would have migrated all the way to the outside of the body [Johnson, 2007, p. 15]. When life forms are compared in this case no clear progression evolutionary pattern is found.

The sperm are stored in the vas deferens in a state of suspended animation due to the close to neutral pH caused in part by the carbon dioxide released by the sperm. The vaginal environment is normally acidic (pH 3.5 to 4.0) and must be neutralized by the semen's buffers. Poorly protected sperm in the seminal fluids suffer significant attrition before being deposited in the seminal coagulum located near the female cervix.[24]

---

[21]Setchell, 2006, p. 81.
[22]Sodera, 2009, p. 110.
[23]Johnson, 2007, p. 15.
[24]Roberts and Jarvi, 2009.

The Kinking Problem

Most body 'tubes' and ducts, including the vas deferens, are soft and flexible. This design is required so that the contents can be effectively transported via peristaltic waves. When bent, the cross-sectional area decreases at the bend, which at the least increases a tube's resistance to the flow of liquids. When the vas deferens bends significantly, the thick vas deferens muscle wall closes off the lumen through which the semen must move inside of the vas deferens.

The vas deferens would have exactly the same 'ovalling' problem as other soft tubes if it took the direct route that Dawkins advocates. But the current vas deferens design effectively avoids the bends that would result from more direct routes. Certain soft body tubes, such as the small intestines, have structures analogous to man-made 'elbows' to prevent kinking.

This short vas deferens design would also not work because it must be pulled up and down with the testes movements that are required for thermoregulation. The velocity of the ejaculate would also be impeded by the use of elbow-designed joints. A direct route taken by the vas deferens would cause bending that would not only obstruct semen flow, but would also weaken the peristaltic wave action even if the tube were not entirely closed. Kinking would also interfere with the correct peristaltic contraction and expansion movement at the proper times.

Vas Deferens Length and Sperm Count Requirements

Individuals with at least 20 million sperm/ml, or a total of 50 million per ejaculation, are likely to be fertile. Males with fewer than ten million spermatozoa/ml are likely to be classified as sterile.[25] Most sperm mature in the epididymis and then are stored in the vas deferens. A vas deferens significantly shorter than 45 cm long would likely result in more male infertility because it could both store and carry a significantly smaller

---

[25]Moore and Persaud, 2008, p. 31.

sperm volume. The result would be low ejaculate volume, resulting in a lower total sperm count. Related to the length is the fact that differences in contractile strength, and in the receptors in the vas deferens between the distal and proximal ends, would be significantly less than now existing in normal, healthy males. This would result in a lower propulsion of the semen, thus would adversely affect fertility.

Furthermore, there exist important differences along the length of the vas deferens, such as differences in contractile strength and in the receptors in the distal and proximal ends that indicate its length has clear functions.[26] Furthermore, "expression of Fst288 and Fst315 and follistatin protein levels increased progressively from the testis through to the distal vas deferens" which "indicate a role for activin A within the epididymis, but also that activin A bioactivity may be increasingly inhibited by follistatin distally along the male reproductive tract."[27] Other differences include the response to ATP or noradrenaline existing along its length, which in turn alters the contractility strength. In this situation, ATP is a neurotransmitter acting through P2X channels.[28]

Required Semen Composition

Semen is composed of numerous ingredients that must be added by various glands located along the vas deferens route and the ejaculatory duct urethra path. These include the many substances necessary for sperm movement, maturation, and maintenance.[29] When spermatozoa coming from the testes enters the vas deferens, the secretions from the seminal vesicles, located slightly superior to the prostate, and from other glands, meet and mix. This requires a sufficient length of tubing for the required level of mixing.

---

[26]Winnall, et al., 2013.
[27]Winnall, et al., 2013, p. 570.
[28]Koslov and Andersson, 2013.
[29]Jones and Lopez, 2006, p. 251.

Kinetic Energy Requirements

For the sperm in the vas deferens to be viable, the vas deferens must receive secretions from several accessory organs that not only must be mixed, but also propelled forward with enough force to reach the goal of fertilization. For fluids to travel between points A and B, a sufficient pressure difference between these points is required. A large enough force difference in thermal fluid networks must exist to create the pressure required to push the fluid forward at the required speed.

The vas deferens is a flexible duct with a relatively thick surrounding muscle that produces the required level of peristalsis.[30] It is a powerful muscle for its size and has the highest muscle-to-lumen ratio of any hollow viscous duct in the body. Peristalsis is responsible for propelling the sperm in the vas deferens into the ejaculatory ducts. The length of journey of sperm from the testes to the urethra is estimated to be proportionate to a man swimming across the Atlantic Ocean.

Peristaltic muscular action moves semi-solid or liquid matter from one place to another in many other body areas. A peristaltic wave contracts behind the contents of one section of the tube, and at the same time relaxes on the front side of the fluid, forcing the duct's contents along the inside of the tube. The existing vas deferens length produces the power level required to propel the semen forward at the velocity level required.[31] Because the ejaculatory ducts lack a muscle coat, the "power for ejaculation comes primarily from the smooth muscle of the ductus [vas] deferens and...to a lesser extent, from the smooth muscle of the seminal vesicles and prostate gland."[32]

These muscles are also responsible for creating the necessary pressure rise in the vas deferens to add the required kinetic energy to the fluid via the vas deferens peristaltic waves. The existing vas deferens length allows peristalsis to produce the power required to transport the seminal fluid with a force sufficient to achieve fertilization. Dawkins erroneously

---

[30]Henrikson, et al., 1997, p. 404.
[31]Roberts and Jarri, 2009.
[32]Henrikson, et al., 1997, p. 404.

treated a dynamic system as a static one. The power produced by the vas deferens' peristaltic waves, in addition to the prostate contractions, provides the spermatozoa with the velocity required to properly propel the semen out of the urethra during emission.

Other Structures in the System

The high kinetic energy level produced by the length of the existing vas deferens ensures that the spermatozoa and semen mix sufficiently with the products of the other sex glands, especially those of the seminal vesicles. The highly complex semen constituents are required to perform multiple sequential functions necessary in order to deliver viable spermatozoa to the correct location at the right time.

The glands involved include the seminal vesicles (named because it was incorrectly believed that their function was to store semen), the prostate gland (discussed above), and the bulbourethral (Cowper's) glands that contribute mucus, which serves as a lubricant.[33] The many nutrients required, including fructose and citric acid to keep the sperm alive, are secreted by both the seminal vesicles and the prostate gland.[34] They also secrete anti-bacterial agents, such as zinc, that are necessary to reduce problematic bacterial infections.

The seminal vesicular secretions contribute about 60 percent of the ejaculate, and the prostate about 30 percent. The seminal vesicles contract to expel their secretions just after the vas deferens peristaltic contraction, which also occurs in concert with the prostate contraction to ensure that the male urethra is emptied of most of the sperm during ejaculation.

Another essential ingredient of semen is prostate specific antigen (PSA), produced by prostate gland epithelial cells. PSA breaks down the coagulum, the sperm-entrapping gel composed of semenogelins and fibronectin, resulting in liquefying the semen in the seminal coagulum, allowing sperm to swim freely.

---

[33]Jones and Lopez, 2006, p. 232; Carlson, 1996.
[34]Jones and Lopez, 2006, p. 232.

Small contributions from the other glands enter directly into the prostatic urethra when required. A direct route by the vas deferens would not allow for the effective mixing of these many secretions. The high power of the vas deferens causes a high exit velocity of sperm coming from the vas deferens, which insures that the secretions from the rest of the sexual glands mix properly.

Only after these glands have made their contributions is the semen able to perform its life-giving functions. As noted, the products of these glands must properly mix, and the 45-cm length of the vas deferens enables this mixing, especially in the section where the seminal vesicles are located, posterior-inferior to the prostate. In the distal part of the epididymis and the vas deferens, the spermatozoa consist of only ten percent of the ejaculate, and effective mixing is essential to achieve the sperm's maximum swimming behavior.

The Role of the Prostate

The prostate is also of critical importance in determining the required location of the vas deferens tube. The entire prostate contracts during the sexual response that, in turn, contracts the U-shaped urethra, forcing the semen in the urethra forward. At ejaculation, the bladder neck closes, and only then do the prostate muscles contract. Because the prostate's many small cavernous holes all open directly into its U-shaped opening, prostate contraction helps to effectively propel the semen forward during ejaculation.

Sperm must travel out of the urethra with sufficient force to successfully complete its journey to achieve fertilization. To aid this requirement the urethral segment of the prostatic squeezes to push the fluid out of the prostate during ejaculation.[35] The fact that the normal healthy prostate is composed of about 40 percent smooth muscle tissue, 40 percent connective tissue, and only 20 percent glandular tissue, indicates the muscular function's critical importance.

---

[35]Baggish, 1996, p. 2.

Among the prostate's other functions are the production of secretions that constitute about a third of the seminal fluid volume.[36] The prostate secretions consist of a thin, opalescent liquid that contains a complex soup of compounds, including citric acid, phosphatase, prostaglandin (named after the prostate gland because it was first discovered there), beta glucuronidase, a potent fibrinolysin, spermine, several proteolytic enzymes, and buffers.[37]

The prostate also adds an alkaline mucus (pH 7-8) that serves to aid in neutralizing the acidic male urethra and female reproductive tracts that are required for sperm to survive its long journey from the testicles to the uterine (fallopian) tubes.[38] The alkaline additives added by the prostate and other organs usually raises the pH of the semen to a slightly basic level of around 7.2, which greatly increases sperm motility.[39] Sperm become active only in a mildly alkaline fluid of about pH 7.5 that results from mixing with the alkaline fluids from the prostate and seminal vesicles.[40]

The Complex Ejaculation Process

The relatively complex ejaculation process is divided into several steps, including a sympathetic nervous system response where semen is deposited into the prostatic urethra via the ejaculatory ducts just after peristalsis in the vas deferens ampulla. Then ejaculation, the phase in which the semen is expelled through the external urethral orifice after having traveled through the urethra, occurs.[41]

This last step involves the periurethral muscles (muscles encircling the urethra) performing a series of rhythmic involuntary contractions that expel the semen. The vesicle sphincter at the neck of the bladder closes tightly during this process to prevent urine from contaminating

---

[36]Swanson and Forrest, 1984.
[37]Bloom and Fawcett, 1969.
[38]Bergman, 2008.
[39]Jones and Lopez, 2006, p. 235.
[40]Jones and Lopez, 2006, p. 232.
[41]Moore and Persaud, 2008, p. 28; Jones and Lopez, 2006, p. 113.

the semen and the backwashing of the ejaculate.[42]

Backflow Prevention

One advantage of the existing design is it protects against urine infiltration into the vas deferens. The first protection is the internal urethral sphincter that prevents urine from entering the urethra during an erection.[43] The existing system also helps to prevent urine from entering the vas deferens due to the fact that the vas deferens is a significant distance *above* the entry to the urethra. Thus, gravity can also help prevent urine from entering the vas deferens.

After an ejaculation, one valve relaxes and the other contracts, to prevent backflow. The circular route means an uphill flow is required to cause urine to flow back into the epididymis. Very heavy weightlifting and other very strenuous activity can cause some urine backflow into the testes, but in ordinary activities the sheer distance of the vas deferens uphill around and over the bladder, then down into the epididymis, helps prevent urine from moving into the epididymis, which could cause major health problems, including infections.

The Vas Deferens in Nonhumans

Most all vertebrates use some form of duct system to transfer sperm from the testes to the urethra. Cartilaginous fish and amphibians carry sperm through the archinephric duct, which also is used to help transport urine from the kidneys. The only vertebrates that lack a structure resembling a vas deferens are the jawless fishes, which release sperm directly into their body cavity, then into the surrounding water through a small body opening.[44] In teleosts, a large group of fish that includes all ray-finned fishes, and some higher fish, a distinct sperm duct is separate from the ureters, (often called the vas deferens) is used.

---

[42]Moore and Persaud, 2008, p. 28.
[43]Greenberg, et al., 2017, p. 166.
[44]Romer and Parsons, 1986.

In general, although little research has been completed on how the vas deferens and related structures function in nonhumans, and "the available data remain scanty," in general, according to the research thus far completed, the design is basically similar in all sexually reproducing mammals.[45] Most of the work on mammal vas deferens was done on the rat, and much of the work done on other mammals was related to fertility regulation.[46] Also, much work on comparative vas deferens focused on its histology.[47]

The vas deferens design in birds is also very similar to that of mammals, one difference being most birds are comparatively small compared with the larger mammals and humans. For this reason, the vas deferens would not be nearly as long as in humans. The smaller size requires some major modifications, such as in many birds and reptiles the vas deferens is the main area for sperm storage in males, not the epididymis as is true in many mammals.[48] A major difference, which may explain some design differences, is humans are sexually active for most of the year, whereas most animals are active only during mating season, which is typically very short.

Another difference is, in the male avian reproductive system, sperm remain viable at normal body temperature, thus they do not require many of the structural designs existing in humans, including a scrotum controlled by a cremaster muscle system, to regulate temperature.[49] Conversely, the similarities are many. One example is a detailed analysis of the appearance by light microscope of the vas deferens of the dog, camel, elephant, opossum and monkey finds few differences.[50]

The vas deferens does not take the shortest possible distance from the testes to the penis. Among the numerous reasons documented in the literature for the longer route includes it allows the flexibility required to allow the testes movement toward or away from the body's abdominal cavity to regulate the temperature. It also provides enough length to store

---

[45]Robaire and Hermo,1988, p. 1000.
[46]Chinoy, 1985.
[47]Khan, et al., 2003.
[48]Robaire and Hermo,1988.
[49]Sneddon and Westfall, 1984.
[50]Robaire and Hermo. 1988, p. 1002.

and mix the semen's essential components, and to avoid the kinking problem that occurs when a soft tube bends. These and other reasons demonstrate the fact that the existing length is necessary for the system to properly function. It is documented that the existing design is nothing short of ingenious.

Conclusion

Not only is the vas deferens design necessary for embryonic development, but there exist several critical reasons for its existing design.[51] Most of these reasons are straightforward and well-known to anatomists. Other reasons, including those raised here, involve fluid mechanics, peristalsis, and the length requirement, all of which become clear when studying the vas deferens configuration from an engineering design point of view. The fact is, the direct route would result in a vas deferens so short that it would impede reproduction to the extent that it could be lethal to some species.

If the existing vas deferens was far longer than is necessary, Darwinists predict that over time random genetic variation and selective pressures would be expected to shorten it. If its length is, indeed, dysfunctional, or at least unnecessary for its proper function, there would exist no biological reason for the great length to be consistently conserved. Furthermore, a total absence of viable superior anatomical designs exist in the literature. While it is common to propose fixes for putative inferior design, such as the common problem of back problems in human bodies, the absence of examples that demonstrated "improvements" to healthy normal bodies documents that human anatomy reflects a highly-optimized level of design.

---

[51] Burnstock and Verkhratsky, 2010.

References

Anonymous. 2011. The Case Against Creationism - Part One; Bad Design. February 12. `http://www.thebeautifultruth.org.uk/index.php?option=com_content&view=article&id=33:the-case-against-creationism-part-one-bad-design&catid=3:science&Itemid=5`.

Baggish, Jeff. 1996. *Making the Prostate Therapy Decision*. Los Angeles, CA: Lowell House.

Blechschmidt, E. and Freeman, B., 2004. *The Ontogenetic Basis of Human Anatomy. A Biodynamic Approach to Development from Conception to Birth*, New York: North Atlantic Books.

Bloom, William and Don Fawcett. 1969. *A Textbook of Histology*. Philadelphia, PA: Saunders.

Burnstock, Geoffrey and Alexej Verkhratsky. 2010. Vas Deferens- A Model Used to Establish Sympathetic Cotransmission. *Trends in Pharmacological Sciences*. 31(3)131-139.

Burrows, W. H. and J. P. Quinn. 1937. The Collection of Spermatozoa from the Domestic Fowl and Turkey. *Poultry Science*, 16(1):19-24, January, 1

Carlson, Bruce. 1996. *Patten's Foundations of Embryology. 6th edition*. New York: McGraw Hill.

Chinoy, N. J. 1985. Structure and physiology of mammalian vas deferens in relation to fertility regulation. *Journal of Biosciences*. 7(2)215-221. March.

Chung, Eric and Gerald Brock. 2011. "Cryptorchidism and its impact on male fertility: a state of art review of current literature. *Canadian Urological Association Journal*. 5(3)210-214.

Dawkins, Richard. (2009). *The Greatest Show on Earth: The Evidence for Evolution*. New York: The Free Press.

Greenberg, Jerrold S., Clint E. Bruess, and Sara B. Oswalt. 2017. *Exploring the Dimensions of Human Sexuality*. Burlington, MA: Jones and Bartlett.

Henrikson, Ray. C. Gordon Kaye and Joseph Mazurkiewicz. 1997. *Histol-*

ogy. Third edition. New York: Lippincott Williams & Wilkins.

Johnson, Martin H. 2007. *Essential Reproduction. Sixth Edition*. New York: Wiley-Blackwell.

Jones, Richard and Kristin Lopez. 2006. *Human Reproductive Biology*. New York: Elsevier.

Khan, Aijaz, A., Zaidi, M.T; Faruqi, N.A. 2003. Ductus Deferens-a Comparative Histology in Mammals. *Journal of the Anatomy Society of India*. 52(2):163-165.

Knobil, Ernst, Jimmy D. Neill, Gilbert S. Greenwald, Clement L. Markert, and Donald W. Pfaff. 1993. *The Physiology of Reproduction*. Second Edition. New York: Lippincott Williams & Wilkins.

Koslov and Andersson. 2013. Physiological and pharmacological aspects of the vas deferens-an update. *Frontiers in Pharmacology*.4:101. August 22.

Lychkova, A., Puzikov, M. 2015. A longitudinal and bilateral asymmetry of the serotonin function in the regulation of motor activity of the rabbit vas deferens in vivo. *Labome*; 2:1475.

Moore, Keith L. and T. V. N. Persaud. 200). *The Developing Human. Clinically Oriented Embryology*. Philadelphia, PA: Sanders Elsevier.

Robaire, B. and L. Hermo. 1988. Efferent Ducts, Epididymis, and Vas Deferens: Structure, Functions, and their Regulation. Chapter 23 in *The Physiology of Reproduction*. Edited by E. Knobil and J. Neill et al., New York: Raven Press.

Roberts, M. and Keith Jarvi. (2009). "Steps in the Investigation and Management of Low Semen Volume in the Infertile Man. *Canadian Urology Association Journal*. 3(6): 479-485. December.

Romer, Alfred and Thomas Parsons. 1986. *The Vertebrate Body*. Philadelphia, PA: Sanders.

Setchell, B. P. 2006. The effects of heat on the testes of mammals. *Animal Reproduction*. 3(2):81-91.

Sneddon, P. and D. P. Westfall, 1984. Pharmacological evidence that adenosine triphosphate and noradrenaline are cotransmitters in the guineapig vas deferens. *Journal of Physiology* 347:561-580.

Sodera, Vija. 2009. *One Small Speck to Man*, 2nd ed, Londn: Vija Sodera Productions.

Steers, W. D. 1994. Physiology of the Vas Deferens. *World Journal of Urology*. 12(5):281–285.

Swanson, Janice M. and Katherine A. Forrest (editors). 1984. *Men's Reproductive Health*. New York: Springer. Springer Series: Focus on Men, Volume 3.

Van Niekerk, E. 2012. Vas Deferens—refuting 'Bad Design' Arguments. *Journal of Creation*. 26(3):60-67.

Wechalekar, Harsha, Brian P. Setchell, Eleanor J. Peirce, Mario Ricci, Chris Leigh, William G. Breed. (2010). Whole-body heat exposure induces membrane changes in spermatozoa from the cauda epididymidis of laboratory mice. *Asian Journal of Andrology*. 12(4): 591– 598.

Williams, George C. 1997. *The Pony Fish's Glow: And Other Clues to Plan and Purpose in Nature*. New York: Basic Books.

Winnall, WR, Wu H, Sarraj MA, Rogers PAW, de Kretser DM, Girling JE, Hedger M. (2013). Expression patterns of activin, inhibin and follistatin variants in the adult male mouse reproductive tract suggest important roles in the epididymis and vas deferens. *Reproduction, Fertility and Development*. 2013;25(3):570-80.

Part III

Design of the Nervous System

Chapter 9

The Left Recurrent Laryngeal Nerve

A common evolutionary claim that the human body is poorly designed, which means it is evidence that it was not intelligently designed. Rather, it was cobbled together by the unintelligent process of evolution. One of the most common examples of poor design cited by evolutionists is the left *recurrent laryngeal nerve* (RLN), which controls the larynx (voice box) muscles.

The claim made by Darwinists that evolution is proved because examples of "poor or at least very puzzling design can be accumulated endlessly,."[1] Oxford Professor Richard Dawkins writes one of the best examples of poor design is the left recurrent laryngeal nerve that connects the brain to the larynx helping to control speech. In mammals, this nerve does not take the direct route to connect the brain and throat but rather descends into the chest, loops around the aorta near the heart, then returns back up to the larynx. The result, Dawkins claims, includes the neve is seven times longer than is required. Dawkins' "favourite example of revealingly bad 'design' in animals," is one he learned from one of his college tutors. Professor David Nichols,[2]

---

[1] Walsh, Hoyt, and Miller. 1987. pp. 60-69
[2] Prothero, 2008, pp 37-38.

The Claim Examined

The reason for the claim it was poorly designed is because it travels *down-ward* past the larynx, then around the aorta and, last, *back up* to the larynx. They assume that this longer route is unnecessary, and a much shorter route directly to the larynx would be far more effective. Reasons for the longer route include both developmental and design constraints. Furthermore, the evidence for intelligent design of the existing arrangement is both obvious and compelling.

The main argument is that it is poorly designed because the laryngeal nerve does not take what at *first* appears to be the shortest route from the brain to the larynx, a design true for many other nerves. Examples include the optic nerves, which do not take the shortest route from eyes to the brain's occipital lobe, rather crosses over at the optic chiasm for what are now known to be very good reasons based on optimal design.[3] Likewise, except for the right and left frontal branches of a facial nerve that are supplied by both sides of the brain, the nerves *from the right side of the brain* innervate the *left* side of the body.[4]

Dawkins claims that the RLN is not only an example of poor design but also proof that humans evolved from fish. During the putative mammal evolution area between the head and body was elongated "and the gills disappeared, some of them turning into useful things such as the thyroid and parathyroid glands, and the various other bits and pieces that combine to form the larynx.[5]

He adds certain structures, "including the parts of the larynx, received their blood supply and their nerve connections from the evolutionary descendants of the blood vessels and nerves that, once upon a time, served the gills in orderly sequence."[6]

However, no scientific evidence exists to support gills "turning the thyroid and parathyroid glands. Gill cells are very different from en-

---

[3]Walsh, Hoyt, and Miller. 1987.
[4]Walsh, Hoyt, and Miller. 1987. pp. 60-69.
[5]Dawkins, 2009, p. 360.
[6]Dawkins, 2009, p. 360.

Figure 9.1: The path of the Left Recurrent Laryngeal Nerve

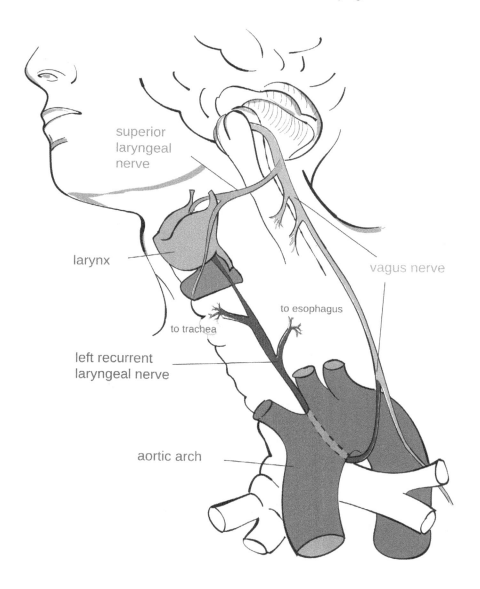

docrine cells. Dawkins concludes, for the reason that as the mammal ancestors

> evolved further and further away form their fish ances-
> tors, nerves and blood vessels found themselves pulled and
> stretched in puzzling directions, which distorted their spatial
> relations one to another. The vertebrate chest and neck became
> a mess, unlike the tidily symmetrical, serial repetitiveness of
> fish gills. And the recurrent laryngeal nerves became more
> than ordinarily exaggerated casualties of this distortion. [6]

Human anatomists know that the vertebrate chest and neck are not a "mess", but a well-designed, complex, fully functional system. Nerve signals that control various bodily operations travel from the brain, down the spine or the cranial nerves, and then branch off to connect to the various organs that they serve. The larynx, located in the neck, and in an embryo the right laryngeal nerve branches off of the vagus nerve in the neck area.[7]

RLN Anatomy

The lingual nerve (LN) is one of the two terminal branches of the posterior division of the large mandibular nerve that controls the larynx.[8] This nerve is also involved with general sensation of anterior two-thirds of the tongue, the sublingual mucosa, the mandibular lingual gingiva and the mouth floor. Two lingual nerves connect to the larynx, the right nerve looping under the right subclavian artery then traveling upwards, and the left loops under the left side of the aortic arch.

Thus, the right and left nerves are not symmetrical. The larynx commonly called the voice box, is located at the top of the neck and produces sound, and protecting the trachea against food aspiration. It also houses the vocal folds, manipulates pitch and volume, which is essential for

---

[7]Mitchell and Sharma, 2009.

[8]Al-Amery, Samah M. 2016. Variation in Lingual Nerve Course: A Human Cadaveric Study. *PLoS One.* 11(9): e0162773. Sep 23.

phonation. The larynx serves numerous functions besides phonation including respiration, airway protection and coordination of swallowing. Several nerves in the larynx control these many major tasks.

The *left* RLN, which supplies the mammalian larynx and epiglottis intrinsic muscles, has for very good reasons an anatomical trajectory very different than one would first expect. In contrast to Prothero's claim, the RLN does not avoid "the direct route between brain and throat and instead descends into the chest, loops around the aorta near the heart, then returns to the larynx."[9]

Rather, the *vagus* nerve trunk travels from the neck down toward the heart, and *then* the recurrent laryngeal nerve *branches* off from the vagus just below the aorta (the large, main artery extending upward from the left ventricle of the heart and extending down the abdomen). Then, as will be detailed, these branches travel upward to innervate several organs, some near where it branches off of the vagus nerve near the aortic arch.[10]

In adults, though, rather than taking the direct route from the vagus nerve to the larynx in the neck area, the *left* vagus nerve travels down the neck into the chest near the heart on its way to provide cholinergic innervation to numerous internal organs. The laryngeal nerve branches off of the vagus nerve in the chest area, then loops *under* the posterior side of the aorta just above the heart and, last, travels *back up* to the larynx. For this reason, it is called the *left recurrent laryngeal nerve*. In contrast, the right laryngeal nerve loops *around* the subclavian artery, and then travels up to the larynx. Of note is the fact that the longer left RLN systems works in perfect harmony with the right RLN, precluding the claim that it is a poor or faulty design.

Reasons for the Design

The most logical reason for this design is due to developmental constraints. Embryologist Professor Erich Blechschmidt wrote that the recurrent laryngeal nerve's seemingly poor design in adults is due to the

---

[9]Prothero, 2008, pp. 37-38.
[10]Sadler, 2010.

"necessary consequences of developmental dynamics and are not to be interpreted ... as historical carryovers" from evolution.[11]

Human-designed devices, such as radios and computers, do not need to function until their assembly is complete. By contrast, during every developmental stage from zygote to adult living organisms must function to a high degree in order to live. The embryo as a whole must also be a fully functioning system in its specific environment during every second of its entire development. For this reason, adult anatomy can be understood *only* in the light of zygote to adult development:

> The pathway for nerve fibers is normally prescribed by the organs-to-be-innervated and is therefore laid down from with-out. We must assume that submicroscopic material (i.e., molecular) movements are decisive for this process; namely, that ordered metabolic movements work in a manner that determines the form of the incipient innervation pattern.[12]

An analogy Blechschmidt uses to elucidate his argument is the course of a river which "cannot be explained on the basis of a knowledge of its sources, its tributaries, or the specific locations of the harbors at its mouth. It is only the total topographical circumstances that determine the river's course."[13]

Due to variations in the topographical landscape of the mammalian body, the "course of the inferior laryngeal nerve is highly variant" and minor anatomic differences are common.[14] This fact has been documented by an anatomy of the right and left recurrent laryngeal nerves study done by a series of neck dissections completed on 90 human cadavers 48 hours following death. The dissections found that the path of these nerves was sometimes different from that shown in the standard medical literature, illustrating Blechschmidt's analogy.[15] One result of the RLN

---

[11]Blechschmidt, 2004, p. 188.

[12]Blechschmidt and Freeman, 2004, p. 8.

[13]Blechschmidt and Freeman, 2004, p. 108.

[14]Sturniolo, G., D'Alia, C., Tonante, A., Gangliano, E., Taranto, F. and Schiavo, M.G., The recurrent laryngeal nerve related to thyroid surgery, *The American Journal of Surgery* 177:487–488, June 1999.

[15]Steinberg, et al., 1986, p. 919.

design is indicates vascular disease because an enlargement of the aorta or subclavian artery caused by an aneurysm may compress the left RLN, causing vocal dysfunction.[16]

Other Developmental Considerations

The human body begins as a sphere called a blastocyst and gradually becomes more elongated as it develops. Some structures, such as the carotid duct, are simply obliterated during development, while others are eliminated and later replaced as the foetus matures. Other structures, including the recurrent laryngeal nerve, are moved downward as development proceeds. The reason for this movement is because neck formation and elongation of the body that occurs during foetal development forces the heart to descend from the cervical (neck) location down into the thoracic (chest) cavity.[17]

As a result, various arteries and other structures must be elongated as organs are moved in such a way so as to remain functional throughout this entire developmental phase. The right recurrent laryngeal nerve is carried radically downward because it is looped under the IV arch which develops into the right subclavian artery, and thus moves down with it.[18]

In cases where the right IV arch is absorbed, the right recurrent laryngeal nerve does not recur, but connects directly into the larynx. The left laryngeal nerve recurs around the ligamentum arteriosum (the VI arch, a small ligament attached to the top surface of the pulmonary trunk and the bottom surface of the aortic arch) on the left side of the aortic arch, thus must move down as the thoracic cavity lengthens. Blechschmidt notes that "No organ could exist that is functionless during its development. This axiom applies to the nervous system. The nervous system achieves its subsequent performances on account of its previous growth functions."[19]

---

[16]Seikel, J.A., King, D.W. and Drumright, D.G., 2005. *Anatomy and Physiology for Speech, Language and Hearing*, 3rd Edition, Thomson Delmar Learning, New York, p. 186, 2005

[17]Sadler, 1990, p. 211.

[18]Schoenwolf, et al., 2009, p. 407.

[19]Blechschmidt, 2004, p. 91.

As a result of the downward movement "of the heart and the disappearance of the various portions of the aortic arches, the course of the recurrent laryngeal nerves becomes different on the right and left sides."[20] Before this shift, these two nerves supply the sixth brachial arches, but as the heart descends the left nerve hooks around the sixth aortic arch, and then ascends again up to the larynx, thus accounting for their recurrent course. These nerves cannot either be obliterated or replaced because the nerve must function during every foetal development stage. As Sadler explains, on the right side

> the distal part of the sixth aortic arch and the fifth aortic arch disappear, the recurrent laryngeal nerve moves up and hooks around the right subclavian artery. On the left, the nerve does not move up, since the distal part of the sixth aortic arch persists as the ligamentum arteriosum.[21]

This movement appears designed to drag the left RLN downward as the abdominal cavity elongates. Sadler also notes that because the

> musculature of the larynx is derived from mesenchyme of the fourth and sixth brachial arches, all laryngeal muscles are innervated by branches of the tenth cranial nerve, the vagus nerve. The superior laryngeal nerve innervates derivatives of the fourth brachial arch, and the recurrent laryngeal nerve derivatives of the sixth branchial arch.[22]

The body must function as a living functional unit, requiring ligaments and internal connections to secure various related structures together while also allowing for both growth and body/organ movement required for the flexibility necessary for normal daily activities. For the laryngeal nerve, the ligamentum arteriosum functions like the hyoid bone to allow movement. Nerves cannot normally be severed during foetal development and then regrown somewhere else, nor can the body sever

---

[20]Sadler, 1990, p. 211.
[21]Sadler, 1990, p. 212.
[22]Sadler, 1990, p. 229.

nerves to allow the movement of existing nerves elsewhere where they reconnect (figure 1).

Other cases exist of one nerve splitting off early and providing direct innervations, and another taking what seems like a circuitous route. One example is the phrenic nerve (c3, c4, c5 fibres), which arises in the neck and descends to the diaphragm. This is a necessary trajectory since the pericardium and diaphragm arise in the *septum transversum* (a thick mass of cranial tissue that gives rise to parts of the thoracic diaphragm and the ventral mesentery of the foregut) in the neck area of the early embryo.

It then migrates caudally (toward the tail bone) as the embryo enlarges by differential growth of the head and thorax areas, taking the nerve with it. A diaphragm could not have evolved step-wise since a partial diaphragm with a defect results in an imperfect chest-abdomen separation, and even a small defect results in herniation of the gut contents into the chest—which either compresses the lungs or results in strangulation of the gut.

For all these reasons Prothero's following claims are both incorrect and poorly considered: "Not only is this design wasteful, but … the bizarre pathway of this nerve makes perfect sense in evolutionary terms. In fish and early mammal embryos, the precursor of the recurrent laryngeal nerve [is] attached to the sixth gill arch, deep in the neck and body region."[23] The RLN similarity exists in all vertebrates, including fish and mammals, due to similar embryo and foetal developmental constraints and similar morphology as an adult. It is not because we evolved from fish as Prothero claims, something which fossil or other evidence does not support.

Other RLN Functions

The recurrent laryngeal nerve branches also serve several other organs, including the upper oesophagus, the trachea, the inferior pharynx, and the circopharyngeus, providing both motor and sensory branches that

---

[23]Prothero, 2008, pp. 37-38.

require its existing design.[24, 25] This arrangement allows the structures *below* the larynx to receive signals slightly sooner than the larynx to prepare them for laryngeal action when this function is imminent.

In addition, "the laryngeal branch splits up into other branches before entering the larynx at different levels."[26] The neuroanatomy of larynx innervation is very complicated and researchers are still trying to work out the specific targets of its nerve branches. The fact also some fibres branch the left RLN that connect to the *cardiac plexus* is also highly indicative of developmental constraints because the nerve must innervate both the larynx, in the neck and the heart in the chest.

A complex issue still being researched is how the incredibly complex nerve-muscle system, consisting of nerve fibres and the laryngeal muscles, arises from the neural crest and dorsal somites respectively in the early embryo and migrates anteriorly into its final position. Claiming that the RLN is poorly designed without understanding how the nerve structure design, its function, and its ultimate origination and connections in the brain developed from embryo to an adult is irresponsible. Thus, the claim that no reason exists for the RLN to loop up the distance from the ligamentum arteriosum is invalid.

The Redundant Pathway Design

Some innervations to the larynx travel directly to it, including the *sensory internal laryngeal nerve* and the *motor external laryngeal nerve*. Two other nerves, the left and right *superior laryngeal nerves*, branch off close to the larynx to provide it with direct innervation. The superior laryngeal nerve branches off the vagus at the middle of the *ganglion nodosum* and receives a branch from the superior cervical ganglion of the sympathetic nervous system.[27]

Aside from the developmental reasons for the RLN circuitous route,

---

[24]Sturniolo, et al., 1999, p. 487.
[25]Armstrong, et al., 1951.pp 532-539
[26]Sturniolo, et al., 1999, p. 487.
[27]Sanders and Mu, 1998.

potential benefits for one of the nerves being slightly longer is that over-lapping sensory and motor innervations exist. Better understanding the laryngeal innervation will help us to understand the reasons for the slightly longer route for a nerve, but a strong hint is provided from the fact that the two nerves regulate different vocal responses.

The Superior Laryngeal Nerve divides into internal and external branches. The external branch controls an internal laryngeal muscle, the cricothyroid muscle[28] and innervates muscles responsible for increasing the voice pitch. The various branches of the Recurrent Laryngeal Nerve innervate muscles responsible for functions including reducing pitch, controlling loudness, and vocal fatigue. The three main branches of the RLN innervate several muscle bundles including the Thyoarytenoid muscle, the posterior Cricoarytenoid muscle, and the lateral Cricoary-tenoid muscle. Damage to the nerves that innervate these muscles affect articulation, and when articulation is impaired speech is perceived as "slurred" or "garbled."[29]

Paralysis of the *superior laryngeal nerve* (the non-circuitous nerve) causes the vocal cords lack their normal tone because they cannot lengthen sufficiently, resulting in difficulty in increasing voice loudness, producing an abnormal high pitch, and resulting in vocal fatigue and an inability to sing. In contrast, paralysis of one or more of the three branches of the *recurrent laryngeal nerve* can result in a weak voice that sounds like Mickey Mouse. In severe cases, paralysis of the vocal cords can result.[30]

One patient, who suffered from a traumatic rupture of his aortic arch in a car accident, required an aortic graft that left him with a damaged left RLN. His articulation was unaffected but his voice was feeble. He speaks perfectly well but he was unable to project his voice properly due to the fact that the laryngeal muscles have multiple innervations and the set *as a unit* control its function (Interview with Dr. Vij Sodera). Another reason why the laryngeal nerve branches, both of which branch off the vagus nerve, are located both above and below the larynx, is because this

---

[28]Bhatnagar, S.C. 2008. *Neuroscience for the Study of Communicative Disorders*. Wolters Kluwer/Lippincott Williams and Wilkins, p. 336, 2008.
[29]Cotton, 2006.
[30]Bhatnagar, S.C. 2008, p 336..

design allows preservation some of function if either one is interrupted. The redundant pathway also provides some back-up in case of damage to one of the nerves.

Last, several studies found that the existing path occupies "a relatively safe position in the tracheoesophageal groove" between the trachea and esophageus[31] that renders it less prone to damage or injury than a more direct route would.[32]

The RLN in Giraffes

The example favoured by those who claim that the RLN is poorly designed is the giraffe. Prothero wrote that the giraffe RLN "traverses the entire neck twice, so it is fifteen feet long (fourteen feet of which are unnecessary!)."[33] Dawkins claims that in humans "the route taken by the recurrent laryngeal nerve represents a detour of perhaps several inches. But in a giraffe, it is beyond a joke—many feet beyond—taking a detour of perhaps 15 feet in a large adult!"[34]

Dawkins added that the "length of the detour taken by the recurrent laryngeal" required a team of anatomists to simultaneously work on different stretches of the nerve to tease out the RLN, which he notes is "a difficult task that had not, as far as we know, been achieved since Richard Owen, the great Victorian anatomist, did it in 1837." The difficulty is due to the fact that the RLN "is very narrow, even thread-like in its recurrent portion," and consequently is difficult to accurately locate

> in the intricate web of membranes and muscles that surround the windpipe. On its downward journey, the nerve (at this point it is bundled in with the larger vagus nerve) passes within inches of the larynx, which is its final destination. Yet it proceeds down the whole length of the neck before turning round and going all the way back up again ... I found my

[31] Duffy, p. 42
[32] Armstrong and Hinton, p. 539
[33] Prothero, 2008, p. 37-38.
[34] Dawkins, 2009, p. 360.

respect for Richard Owen (a bitter foe of Darwin) going up. The creationist Owen, however, failed to draw the obvious conclusion. Any intelligent designer would have hived off the laryngeal nerve on its way down, replacing a journey of many metres by one of a few centimeters.[35]

Many of the same comments that relate to humans discussed above also apply to the giraffe. The giraffe embryo lacks a neck and lengthens, as does the human embryo, only much greater because its neck is designed to be much longer than in humans. Consequently, it moves the RLN down with it as the neck lengthens. Dawkins claims the giraffe's neck lengthened over evolutionary time, even though evidence is lacking for giraffe neck evolution.[36] He concludes without providing evidence that "the cost of the detour—whether economic cost or cost in terms of 'stuttering'—gradually increased" as the neck evolved.[37]

It is true that the significant difference in lengths of the right/left RLN results in the impulses arriving at the giraffe laryngeal muscles at slightly different times, but the impulses along the longer route of the left RLN are adjusted for by the brain in order for the right/left larynx muscles to function smoothly. This indicates forethought in design to compensate for the developmental constraints resulting in right/left RLN length differences. Although giraffes do not talk, thus cannot stutter as Dawkins claims, and make very little noise, they do have a larynx that must function.

Dawkins then claims that, as the "giraffe's neck began to approach its present impressive length, the *total* cost of the detour might have begun to approach the point where—hypothetically—a mutant individual would survive better if its descending laryngeal nerve fibres hived themselves off from the vagus bundle and hopped across the tiny gap to the larynx.[38] Dawkins recognizes that this mutation is not feasible because the mutation required

to achieve this "hop across" would have to have consti-

---

[35]Dawkins, 2009, pp. 361-362.
[36]Bergman, 2002.
[37]Dawkins, 2009, p. 363.
[38]Dawkins, 2009, pp. 363-364; Dawkins, 2009, p. 363.

tuted a major change—upheaval even—in embryonic development. Very probably, the necessary mutation would never happen to arise anyway. Even if it did, it might well have disadvantages—inevitable in any major upheaval during the course of a sensitive and delicate process. And even if these disadvantages might eventually have been outweighed by the advantages of bypassing the detour, the *marginal* cost of each millimeter of *increased* detour *compared with the existing detour* is slight.[39]

He adds:

Even if a "back to the drawing board" solution would be a better idea if it could be achieved, the competing alternative was just a tiny increase over the existing detour, and the *marginal* cost of its tiny increase would have been small. Smaller, I am conjecturing, than the cost of the "major upheaval" required to bring about the more elegant solution."[40]

Dawkins' conclusion is an excellent and valid argument against evolution-by-mutations theory. He irresponsibly concludes that his main point is "the recurrent laryngeal nerve in any mammal is good evidence against a designer" but

is exactly the kind of thing we expect from evolution by natural selection, and exactly the kind of thing we do *not* expect from any kind of intelligent designer ... If this were designed, nobody could seriously deny that the designer had made a bad error. But, just as with the recurrent laryngeal nerve, all becomes clear when we look at evolutionary history.[41]

---

[39]Dawkins, 2009, pp. 363-364.
[40]Dawkins, 2009, pp. 363-364.
[41]Dawkins, 2009, pp. 363-364.

Conclusions

The left recurrent laryngeal nerve is not poorly designed as claimed by many Darwinists, but rather is evidence of both good and intelligent design. No evidence exists that the design causes any disadvantage, and much evidence exists in favour of the conclusion that the existing design results from developmental constraints and also serves to fine-tune laryngeal functions. The arguments presented by evolutionists are incorrect and have discouraged research into the specific reasons for the existing design.

The constraints resulting from foetal development are actually similar to the evolutionists' argument. The only difference is that there are two different developments involved, 1) ontogeny, which is referred to in this chapter, and 2) phylogeny, which is referred to by evolutionists. The evolutionary "proof" becomes worthless when an equally valid explanation exists based on the individual's historical development, the ontogeny.

To argue that the RLN is poorly designed is to imply that God should have designed different embryo development trajectories for all of the structures involved to avoid looping the RLN around the aorta. One who asserts that the RLN is a poor design assumes that a better design exists, a claim that cannot be asserted unless an alternative embryonic design from fertilized ovum to foetus (including all the incalculable molecular gradients, triggers, cascades and anatomical twists and tucks) can be proposed that would document an improved design. Lacking this information, the "poor design" claim is an "evolution of the gaps" explanation. Other alternatives to embryonic design or developmental pathways would likely result in its own unique set of new constraints producing more problems, also giving the impression of poor design.

References

Armstrong, W.G. and Hinton, J.W., Multiple divisions of the recurrent laryngeal nerve, *AMA Archives of Surgery* 62(4): 532–539, 1951.

Bergman, Jerry, The giraffe's neck: another icon of evolution falls, *TJ (Journal of Creation)* 16(1):120–127, 2002.

Bhatnagar, S.C. *Neuroscience for the Study of Communicative Disorders.* Wolters Kluwer/Lippincott Williams and Wilkins, p. 336, 2008.

Blechschmidt, E. and Freeman, B., *The Ontogenetic Basis of Human Anatomy. A Biodynamic Approach to Development from Conception to Birth,* North Atlantic Books, New York, p. 188, 2004.

Colton, R.H., J.K. Casper, R. Leonard,S. Thibeault, and R. Kelley. *Unerstanding Voice Problems: A Physiological Perspective for Diagnosis and Treatmen. Third Editiont.* Lippincott Williams and Wilkins, Philadelphia, PA, 2006.

Dawkins, Richard. *An Appetite for Wonder: The Making of a Scientist.* New York: Ecco. Pp159-160, 2013.

Dawkins, R., 2009. *The Greatest Show on Earth: The Evidence for Evolution,* Free Press, New York, p. 360.

Duffy, J.R. *Motor Speech Disorders: Substrates, Differential Diagnosis, and Management. Second Edition.* Elsevier Mosby, p. 42, 2005.

Mitchell, B. and Sharma, R., *Embryology,* Churchill Livingstone Elsevier, Philadelphia, PA, 2009.

Prothero, D., *Evolution: What the Fossils say and Why it Matters,* Columbia University Press, New York, pp. 37–38, 2008.

Sadler, T.W., *Langman's Medical Embryology.* 6th Edition, Williams & Wilkins, Philadelphia, PA, p. 211, 1990.

Sanders, I. and Mu, L., Anatomy of the human internal superior laryngeal nerve, *The Anatomical Record* 252:646–656, 1998.

Schoenwolf, G.C., Bleyl, S.B., Brauer, P.R. and Francis-West, P.H., *Larsen's Human Embryology.* Churchill Livingstone, Philadelphia, PA, p. 407, 2009.

Seikel, J.A., D.W. King, and D.G. Drumright. *Anatomy and Physiology for Speech, Language and Hearing. Third Edition.* Thomson Delmar

Learning, United States, p. 186, 2005.

Steinberg, J.L., Khane, G.J., Fernanades, C.M.C. and Nel, J.P., Anatomy of the recurrent laryngeal nerve: A redescription, *The Journal of Laryngology and Otology* 100, p. 919, August 1986.

Sturniolo, G., D'Alia, C., Tonante, A., Gangliano, E., Taranto, F, and Schiavo, M.G., The recurrent laryngeal nerve related to thyroid surgery, *The American Journal of Surgery* 177, pp. 487–488, June 1999.

Walsh, Frank Burton, William Fletcher Hoyt, and Neil R. Miller. 1997. Walsh and Hoyt's Clinical neuro-ophthalmology, Volume 4. Baltimore, MD: Williams & Wilkins. Chapter 4: Anatomy and Physiology of the Optic Chiasm.

Chapter 10

The Inverted Retina

Introduction

It is often claimed that the human retina is poorly designed because light must travel *through* nerves and blood vessels to reach the photoreceptor cells, which are located *behind* the eye's wiring. Many specific reasons exist for this so-called backward placement of the photoreceptors. A major one is that it allows close association between the rods and cones and the pigment epithelium required to maintain the photoreceptors. It is also essential for both the development and the normal retina functions. Both the rods and cones must physically interact with retinal pigment epithelial cells, which provide nutrients to the retina, recycles photopigments, and creates an opaque layer to absorb excess light.

The Alleged Problem

One of the most common examples of putative poor design in both the popular and scientific literature is the mammalian retina. The retina is the thin, light-sensitive organ located at the back of the eyeball. The claim is made that the vertebrate eye is functionally suboptimal because the retina

137

photoreceptors are oriented *away* from incoming light.[1] Shermer claimed that anatomy of the human eye shows it is not intelligently designed because it is

> built upside down and backward, with photons of light having to travel through the cornea, lens, aqueous fluid, blood vessels, ganglion cells, amacrine cells, horizontal cells, and bipolar cells, before reaching the light-sensitive rods and cones that will transduce the light signal into neural impulses.[2]

Oxford Professor Richard Dawkins considers this an example of poor design because he concludes that an

> engineer would naturally assume that the photocells would point towards the light, with their wires leading backwards towards the brain. He would laugh at any suggestion that the photocells might point away from the light, with their wires departing on the side *nearest* the light. Yet this is exactly what happens in all vertebrate retinas. Each photocell is, in effect, wired in backwards, with its wire sticking out on the side nearest the light. The wire has to travel over the surface of the retina, to a point where it dives through a hole in the retina (the so-called 'blind spot') to join the optic nerve.[3]

He admits light entering the eye

> instead of being granted an unrestricted passage to the photocells, has to pass through a forest of connecting wires, presumably suffering at least some attenuation and distortion (actually probably not much but, still, it is the principle of the thing that would offend any tidy-minded engineer!).[4]

Williams claimed the retina is not just an example, but one of the *best* examples of "poor design" in vertebrates that proves a "blind watchmaker" created life:

---

[1] Ayoub, 1996, p. 19.
[2] Shermer 2005, p. 186.
[3] Dawkins, 1986, p. 93.
[4] Dawkins, 1986, p. 93.

Every organism shows features that are functionally arbitrary or even maladaptive. . . . *My chosen classic is the vertebrate eye.* It was used by Paley as a particularly forceful part of his theological argument from design. As he claimed, the eye is surely a superbly fashioned optical instrument. It is also something else, a superb example of maladaptive historical legacy. . . . Unfortunately for Paley's argument, the retina is upside down. The rods and cones are the bottom layer, and light reaches them only after passing through the nerves and blood vessels.[5]

Williams admitted that the vertebrate eye still functions extremely well in spite of the backward retina, but argued that this does not negate the "fact of maladaptive design, however minimal in effect," which disproves "Paley's argument that the eye shows intelligent prior planning."[6] Barash and Barash even claimed that the human

eye, for all its effectiveness, has a major design flaw. The optic nerve, after accumulating information from our rods and cones, does not travel directly inward from the retina toward the brain as any minimally competent engineer would demand. Rather, for a variety of reasons related to the accidents of evolutionary history plus the vagaries of embryonic development, optic-nerve fibers first head away from the brain, into the eye cavity, before coalescing and finally turning 180 degrees, exiting at last through a hole in the retina and going to the brain's optic regions.[7]

Tuff's University Professor Daniel Dennett argued that, although the eye's design is brilliant,

it betrays its origin with a tell-tale flaw: the retina is inside out. The nerve fibers that carry the signals from the eye's rods

---

[5]Williams, *emphasis mine*, 1992, p. 72.
[6]Williams, 1992, p. 73.
[7]Barash and Barash, 2000, p. 296.

and cones (which sense light and color) lie on top of them, and have to plunge through a large hole in the retina to get to the brain, creating the blind spot. No intelligent designer would put such a clumsy arrangement in a camcorder, and this is just one of hundreds of accidents frozen in evolutionary history that confirm the mindlessness of the historical process.[8]

After noting that the backward retina is a "classic" example of the "stupid features which support the idea that they are the result of evolution by natural selection." Frymire concluded that the inverted retina "results in an absurd situation in which the light has to travel through blood vessels and nerves before it reaches the rods and cones."[9] Diamond added that, of all of our anatomy features

> none is more often cited by creationists in their attempts to refute natural selection than the human eye. In their opinion, so complex and perfect an organ could only have been created by design. Yet while it's true that our eyes serve us well, we would see even better if they weren't flawed by some bad design. Like other cells in our bodies, the retina's photoreceptor cells are linked to a network of blood vessels and nerves. However, the vessels and nerves aren't located behind the photoreceptors, where any sensible engineer would have placed them, but out in front of them, where they screen some of the incoming light.[10]

He adds by contrast

> the eyes of the lowly squid, with the nerves artfully hidden behind the photoreceptors, are an example of design perfection. If the Creator had indeed lavished his best design on the creature he shaped in his own image, creationists would surely have to conclude that God is really a squid.[11]

---

[8]Dennett, 2005.

[9]Frymire, 2000, p. 36.

[10]Diamond, 1985, p. 91.

[11]Diamond, 1985, p. 91.

Kenneth Miller claimed that a prime example of "poor design" is the fact that in the human eye light has to travel through the neuron layers before it reaches the retina photoreceptors. He argued that this design provides clear evidence that the eye evolved by mutations and natural selection and was not designed. An intelligent designer, he maintained, would not have placed "the neural wiring of the retina on the side facing incoming light. This arrangement scatters the light, making our vision less detailed than it might be."[12] Thwaites argued that the inverted retina problem hits at the *core* of the design argument, historically a major basis of theism because the "vertebrate eye shows poor design when compared to the eye evolved by the cephalopods" because vertebrate's see everything

> through the nerves and blood vessels of the retina since the photosensitive elements of the retina are on the far side of the retina away from the light source. Clearly the cephalopod solution to retinal structure is more logical, for they have the photosensitive elements of the retina facing the light. Certainly the creationists need to explain why we got the inferior design. I had thought that people were supposed to be the Creator's chosen organism.[13]

Williams added that "our eyes, and those of all other vertebrates, have the functionally stupid upside-down orientation of the retina" and that the "functionally sensible arrangement is in fact what is found in the eye of a squid and other mollusks."[14]

The so-called inversion of the retina is considered a suboptimal design primarily because of its simplistic comparison with a camera. Diamond argued that placing the rods and cones at the bottom layer and requiring light to pass through the nerves and blood vessels is the opposite of how an engineer would have designed the eye, and "a camera designer who committed such a blunder would be fired immediately."[15]

And Edinger concluded that the "vertebrate eye is like a camera with

---

[12]Miller, 1999, p. 10.
[13]Thwaites, 1982, p. 210.
[14]Williams, 1997, pp. 9-10.
[15]Diamond, 1985, p. 91.

the film loaded backward . . . if an engineer at Nikon designed a camera like that, he would be fired."[16] This conclusion is based not only on the assumption that placing nerves and blood vessels in *front* of the retina reduces the retina's overall effectiveness, but that another design would be, as a whole, superior. An evaluation of this argument reveals it is not only naive, but also grossly erroneous.

Verted and Inverted Eyes

Research has clearly shown why the human retina *must* have an "inverted" design, forcing the incoming light to travel *through* the front of the retina to reach the photoreceptors. The opposite placement (where the photoreceptors face the *front* of the eye) is a "verted" design. Verted eyes are wired so that the photoreceptors face *toward* the light and the nerves are placed *behind* the photoreceptor layer.[17]

Most invertebrates and the pineal or dorsal eyes of lower vertebrates use the verted eye design, and most vertebrates (including mammals, birds, amphibians, and fish) use the inverted design. Most verted eye designs are very simple, although a few, such as the cephalopod eye (squids and octopi), are almost as complex as the vertebrate eye.[18] Even the better verted eyes are still "overall quite inferior to the vertebrate eye," a conclusion usually determined by measuring performance of some task in response to visual stimuli.[19] Why this is true will now be discussed.

Advantages of the Inverted Design

One major advantage of the existing design is the "inverted retina actually is a superior space-saving solution" important for small sized eyes.[20]

---

[16]Edinger, 1997, p. 761.
[17]Miller, 1994, p. 30.
[18]Abbott, *et al.*, 1995.
[19]Hamilton, 1985, p. 60.
[20]Kröger and Biehlmaier. 2009, p. 2318.

They studied small fish eyes, and identified a "considerable functional advantage of the inverted retina" due to the stringent space-saving demands of small vertebrates which require small eyes. The refractive power of a fishes thin cornea is very small. Given the high refractive index of sea water and aqueous humor on both sides of the thin cornea, a thick, spherical and optically powerful crystalline lens is required to achieve effective focusing. Focusing requires a significant distance between the lens and the photosensitive retina, and the inverted eye design achieves these requirements and the verted eye does not. The reason is, the space between the lens and the photosensitive layer is almost completely filled with retinal cells, consequently leaving very little room or need for vitreous humor to maintain the required eyeball shape.

The Cephalopod Visual System

The most advanced invertebrate eye known today is that used by certain cephalopods, but the most advanced eye may actually be the extinct trilobite.[21] The cephalopod visual system is not completely understood, both because we have only fossil remnants, and understanding its design is not a funding priority—as is research related to cancer or heart disease. It is known that the major anatomical difference between the human eye and the advanced cephalopod eye, such as the octopus, is the retina, which is not only verted, but also lacks the most sensitive part of the retina, the *fovea centralis*.[22]

The sensitivity of the existing human inverted design is so great that a single photon is able to elicit an electrical response.[23] Consequently, functional sensitivity of the inverted retina could not be significantly improved:

> Neurobiologists have yet to determine how such a negative system of operation might be adaptive, but they marvel over the acute sensitivity possible in rod cells. Apparently rod cells

---

[21]Bergman, 2007.
[22]Land and Nilsson, 2005, p. 64.
[23]Baylor, 1979.

are excellent amplifiers. A single photon (unit of light) can produce a detectable electrical signal in the retina, and the human brain can actually "see" a cluster of five photons—a small point of light, indeed.[24]

Greater sensitivity than this single-photon threshold, if it was possible, might actually result in poorer vision due to sensory overload. In a similar fashion, William's syndrome patients have superior hearing compared to those with average hearing, allowing them to hear a faint whisper; however, this sensitivity causes them serious sensory overload problems, such as in dealing with loud noises like thunder which actually causes them physical pain.

In contrast to the claims of Dawkins, et al., no evidence exists that even the most advanced verted cephalopod eye is superior to the inverted eye. Physiologically, the verted cephalopod retina is simpler compared to the inverted vertebrate retina. An example is there are "no equivalents of the amacrine, bipolar or ganglion cells in the cephalopod retina."[25] The optic lobes, located behind the eyeball in cephalopods, must assume many of the image processing functions that occur in the vertebrate inverted retina. As an underwater animal that usually lives on the ocean bottom, eyes are designed to detect motion, not detail, as is true of human eyes. It must also maximize its utilization of light, since the ocean usually has little or no light at lower depths. The cephalopod eye

undoubtedly forms an image, but the animal's visual perception is certainly quite different from that of man, which is greatly dependent upon interpretation by the brain. The cephalopod optic connections appear to be especially adapted for analyzing vertical and horizontal projections of objects in the visual field.[26]

As Ayoub asked, would "hundreds of thousands of vertebrate species—in a great variety of terrestrial, marine and aerial

---

[24]Ferl and Wallace, 1996, p. 611.
[25]Wells, 1978, p. 150.
[26]Barnes, 1980, p. 454.

environments—really see better with a visual system used by a handful of exclusive marine vertebrates? In the absence of any rigorous comparative evidence all claims that the cephalopod retina is functionally superior to the vertebrate retina remain entirely conjectural."[27]

The cephalopod visual system is designed very differently from the inverted eye in other ways to enable them to function in their water world. Most cephalopods including octopi have only one visual pigment and are thus color-blind.[28] Pechenik indicated that, although cephalopods can perceive shape, light intensity, and texture, they lack many of the advantages of an inverted retina such as the ability to perceive small details.[29] The maximum resolvable spatial frequency in cycles per radian is 4,175 for humans and only 2,632 for octopi.[30] Their photoreceptor cell population is composed entirely of rods, which contain a "mere" 20 million retina receptor cells compared to 126 million in humans.[31]

Their rod outer segments contain rhodopsin pigment with a maximum absorption in the blue-green part of the spectrum (475 nanometers (nm)), which is the predominant color in their environment. In cephalopods, photons change the rhodopsin to metarhodopsin, and no further breakdown or bleaching occurs.[32] A second octopus retina pigment, retinochrome, has an absorption maximum of 490 nm, which is more sensitive to dim light.[33] Humans have one rod type and three cone types. One cone type has a broad peak light frequency of 440 nm (blue), another type 540 nm (green), and the third type 570 nm (red).[34]

The squid's visual system must function in an aqueous medium. Water acts as a filter, and, as a result, the light is of a much lower intensity. Consequently, a squid's vision sensitivity is for shorter wavelengths (below around 400 nm) than a human's, which is from 400 to 700 nm.[35] In bright light the cephalopod's pupils become thin and slit-shaped and are

---

[27] Ayoub, 1996, p. 20.
[28] Land and Nilsson, 2005, p. 64.
[29] Pechenik, 1991, p. 312.
[30] Land and Nilsson, 2005, p. 38.
[31] Young, 1971, p. 441.
[32] Wells, 1978, p. 145.
[33] Wells, 1978, p. 146.
[34] Stoltzmann, 2006.
[35] Peet, 1999, p. 4.

held in a horizontal position by a statocyst, an organ that uses gravity to determine the horizontal.[36] Their visual process is very similar to that of reptiles and insects. A 'photograph' of the recorded image is not traced on the retina as occurs in humans, but rather cephalopods respond to light and color variations of a moving object.[37]

Importantly, the octopus responds to certain motions of non-food objects as if they were prey, but will not react to their normal food-objects if motionless.[38] The importance of motion supports the observation that the octopus eye actually functions as a simple "compound eye with a single lens," because each receptor cell is surrounded by photo-pigment containing microvilli that form a rhabdomeric structure like a compound lens.[39] Each facet in a compound eye is either on or off, and object movement produces a change in the on and off pattern—similar to the manner in which a series of light bulbs produces the illusion of movement by changing the on and off patterns.

Our ignorance about the function of major parts of the cephalopod visual system, such as the optic lobe, prevents researchers from completing a more detailed analysis of cephalopod vision. No evidence exists of how the basic eye types could have evolved from a putative primitive type, in part because neither transitional forms, nor plausible hypothetical intermediate forms exist. The essential difference between vertebrate and invertebrate eyes is the vertebrate eye photoreceptors which face

> outwards towards the choroid, whereas in invertebrates they mostly point inwards towards the lens. But for that obstacle we should have been deluged with theories on the original evolution of the vertebrate eye from the invertebrate. As it is, vertebrate visual origins have to be approached with great caution, and . . . [t]here is nothing indisputable which can be used to explain the origins of the vertebrate eye from an invertebrate organ.[40]

---

[36] Young, 1971.
[37] Grzimek, 1971, p. 191.
[38] Spigel, 1965, p. 126.
[39] Budelmann, 1994, p. 15.
[40] Prince, 1956, pp. 334, 348.

All known animals have either verted or inverted retina eyes, and no evidence exists of transitional forms. Invertebrate eyes use either some type of a lens-based eye, such as cephalopods, or a compound eye as used in trilobites and insects today. All known vertebrates have inverted eyes, and there are no known intermediates between the two. In short, the claim that the cephalopod eye is superior to ours is false, and claims that it is are irresponsible.

Rod and Cone Functions in Vertebrates

The *rods* and *cones* are photoreceptor cells located in the retina used to transduce light into electrical signals. Black and white transduction occurs in the rod-shaped receptors, and color transduction occurs largely in the cone-shaped receptors.[41] The inverted retina vision system requires light to first pass through the cornea, then through the anterior chamber filled with aqueous fluid and, last, the lens and the *vitreous humor*.

Before reaching the retina itself, the light must, in addition, pass through the inner retina's cell layers which contain a dense array of neural processing cells, and on past the rods and cones until it reaches the posterior (distal) end of these cells—wherein lie the outer cell segments. The outer cell segments contain the photoreceptors, light-sensitive structures including *photopigments*, where the transduction of light into receptor potentials occurs.

The photopigment family of proteins undergo physical changes when they absorb light energy. The principal photopigment, opsin glycoprotein, is a derivative of *retinal*, a modified vitamin A molecule. Rods contain a single photopigment type called rhodopsin (rhodo meaning *rose* and *opsis* meaning *vision*). The cones contain one of three different kinds of photopigments: iodopsins, namely *erythrolabe* (most sensitive to red), *chlorolabe* (most sensitive to green) and *cyanolabe* (most sensitive to blue).[42]

Vision functions by light causing changes in the retina photopigments

[41]Ryan, 1994.
[42]Shier, et al., 1999, p. 482.

Figure 10.1: The Main Parts of the Human Eye

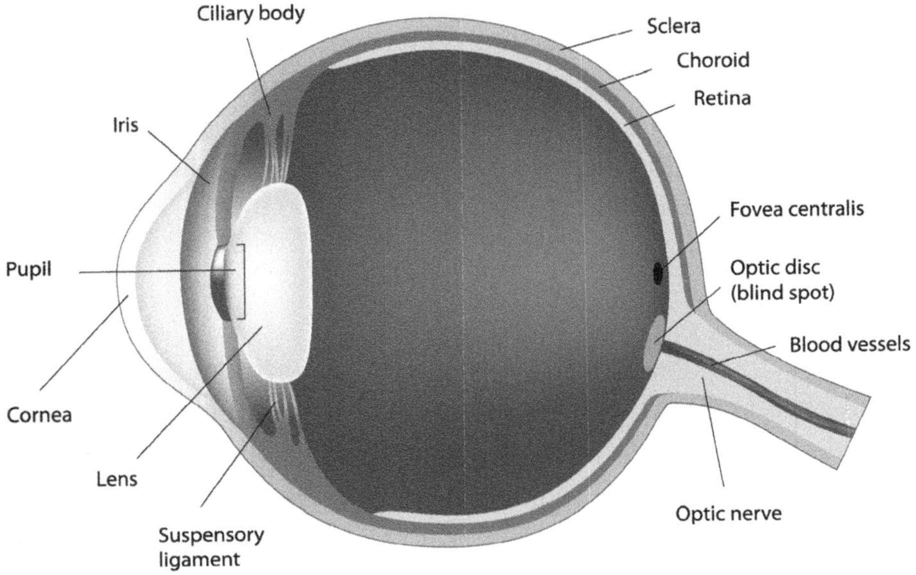

Image Credit: Alila Medical Media / Shutterstock.com

molecule. The molecule has a bent shape (*cis-retinal*) in darkness, and when it absorbs light, isomerization occurs, causing the molecule to form the "straight" form (*trans-retinal*). This causes several unstable intermediate chemicals to form, and, after about a minute, the *trans*-retinal form completely separates from opsin, causing the photopigment to appear colorless (for this reason the process is called bleaching). In order for the rods and cones to again function for vision, retinal must be converted from the *trans* back to the *cis* form. This resynthesis process, called regeneration, requires the retina pigment epithelium (RPE) cells be located *next* to the rod and cone outer segments.

An average of five minutes is required for the rhodopsin regeneration in rods, compared to 1.5 minutes for iodopsin regeneration in cones.[43] Excessive light causes blindness in the affected rods and cones until this regeneration process occurs. This is shown by the temporary blindness

---

[43]Tortora and Grabowski, 1996, p. 468.

Figure 10.2: An Exploded Illustration of the Human Eye

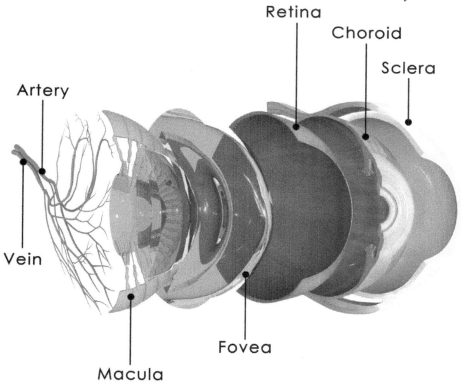

Note the light enters on the left side of the diagram thus must pass through the blood circulation system shown on the far left.

Image Credit: sciencepics / Shutterstock.com

that occurs after watching a bright light flash from a camera strobe light.[44]

Light stimulates rods and cones to release neurotransmitters that induce graded, local potentials in both bipolar and horizontal cells. It is by this means that rod and cone outer segments transduce light into electrical signals. The signals are then carried by the central nervous system neurons to bipolar cells that, in turn, synapse onto the ganglion cells, then to the lateral geniculate body of the thalamus and, last, to the occipital region of the brain stem where the information is organized into a useful

---

[44]Snell and Lemp, 1989.

visual image.[45]

The Retinal Pigment Epithelium

One of the many reasons for the inverted design is that behind the photoreceptors lies a multifunctional and indispensable structure, the *retinal pigment epithelium* (RPE).[46] RPE is a single-cell-thick tissue layer consisting of relatively uniform polygonal shaped cells whose apical end is covered with dense microvilli and basal membrane infoldings. Posterior to the RPE is the vascular choroid layer, and posterior to it is the connective tissue known as the sclera. The RPE touches the extremities of both the rod and the cone photoreceptors, and the RPE microvilli interdigitate with their sides.[47]

The photoreceptors (rods and cones) must face *away* from the front of the eye in order to be in close contact with the vascular choroid, which supplies the photoreceptors with nutrients and oxygen. This arrangement also allows a steady stream of the vital molecule retinal to flow to the rods and cones, without which vision would be impossible.[48] The verted design, on the other hand, places the photoreceptors *away* from their source of nutrition, oxygen, and retinal.

This design would fail because of the enormous amount of energy the rods and cones necessary for the high metabolism required to function. In addition, due to phototoxicity damage from light, the rods and cones must completely replace themselves approximately every seven days. Seemingly simple in appearance, the RPE has "a complex structural and functional polarity that allows them to perform highly specialized roles."[49] One of their major functions is to recycle the used retinal from the photoreceptors.

Vision depends on isomerization of 11-*cis*-retinal to 11-*trans*-retinal in the rods and cones outer segments. Each light photon striking a pho-

---

[45]Stoltzmann, 2006, p. 4.
[46]Martínez-Morales, 2004, p. 766.
[47]Steinberg and Wood, 1994, p. 39.
[48]Kolb, 2003, p. 28.
[49]Hewitt and Adler, 1994, p. 58.

Figure 10.3: Illustration of Human Retina with Rod and Cone Cells

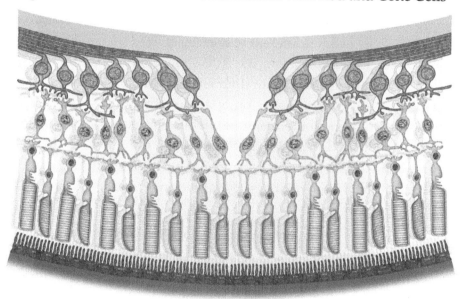

Image Credit: ilusmedical / Shutterstock.com

toreceptor isomerizes retinal, and billions of photons can strike the retina at any one second. The RPE constantly restores the chromophore to *cis*-retinal from its *trans* configuration caused by photostimulation, permitting visual pigment synthesis and regeneration.[50] The 11-*cis*-retinal must also be regularly replaced to maintain the cycle, a task for which the RPE is critical.[51] The RPE manufactures retinal isomerase and other enzymes and stores large quantities of vitamin A that is required to regenerate retinal.

Since RPE cells use enormous amounts of energy and nutrients, they must be in intimate contact with *both* the photoreceptors and the blood supply (in this case the choroid) to carry out these critical functions.[52] Research on the eyes of different species has found that, although major differences exist among them, the RPE shows "little variation."[53] The small RPE variations are due to differences in the retina structure, indicating its critical role in the vision of all vertebrates. One study found retinol isomerase in all the major vertebrates tested and was lacking in all three cephalopods tested.[54] Reciprocal flow of retinoids between the retina and the site of isomerase action in the RPE is a feature common to the visual cycle in all vertebrates.[55] Furthermore,

> the pigment epithelium, which is closely associated with the disk ends of the rods and cones, also provides them with nutrients for making new disks. The epithelium gets its nutrients from the rich blood supply in the choroid layer next to it. In order for the pigment epithelium to function properly, it needs this blood supply. To put both the pigment epithelium and its choroid blood supply on the inside of the eye, between the light source and the light-sensitive rods and cones, would severely disrupt the visual process.[56]

---

[50] Dowling, 1987, p. 198.
[51] Hewitt and Adler, 1994.
[52] Marshall, 1996.
[53] Kuwabara, 1994, p. 58.
[54] Bridges, 1989.
[55] Bridges, 1989, p. 1711.
[56] Roth, 1998, p. 109.

The RPE's Phagocytic Role

A major role of the RPE is to recycle the used rod and cone outer segment membranes, the cone portion closest to the RPE. When the eyes are open, the photoreceptors and RPE continuously absorb an enormous amount of light. This light is converted largely into heat, requiring a very effective cooling system. The choroidal blood supply directly behind the RPE carries away, not only this heat, but also the relatively large amounts of waste products produced by the high level of rod and cone metabolism. Which compounds are allowed to pass through this area are determined by basal membrane receptors.

Cones usually contain from 1,000 to 1,200 discs, and rods from 700 to 1,000. The enormous amount of outer segment activity requires continual replacement of these discs.[57] As the outer segment lengthens from its base, the oldest membrane, which is at the distal end, is shed in segments of one to three discs at a time. Those that are sloughed off are phagocytized by enzymes stored in RPE lysosomes and its components recycled.[58]

The RPE phagocytizes about ten percent of the outer segment discs of normal rod photoreceptors located at its apex, and renews the same amount daily.[59] To replace those segments that are lost, new outer segment membranes are continually being produced at the outer photoreceptor segment base. In short, photoreceptor outer segments are renewed at "an astonishingly rapid pace."[60]

After the RPE breaks down the ingested material, the free radicals and superoxides produced must be neutralized by superoxide dismutase, peroxidase, and other enzymes.[61] This process is continuous, effectively maintaining the photoreceptor's high sensitivity.[62] Furthermore, the RPE carries out several critical functions

for the normal operation of the visual system. One of these

---

[57]Bok, 1994.
[58]Tortora and Grabowski, 1996, p. 467.
[59]Benson, 1996.
[60]Tortora and Grabowski, 1996, p. 467.
[61]Hewitt and Adler, 1994, p. 60.
[62]Benson, 1996.

important roles, appreciated for about a decade, is the phagocytosis of rod outer segment debris. This scavenging activity
goes on daily at an impressive rate in the normal retina. It can
be accelerated to extraordinary levels when outer segments
are damaged. Disruption of this phagocytic function may un-
derlie a variety of clinical disorders, some of which result in
blindness.[63]

RPE microvilli that interdigitate the outer segments surround the photoreceptor so as to effectively carry out their phagocytic and recycling
role.[64]

Nutrient Role of the RPE

The RPE selectively transports nutrients from the choroidal circulation
system to both the photoreceptors and retinal cells. The RPE also helps
maintain water and ion flow between the neural retina and the choroid,
protects against free radical damage, and regulates retinoid metabolism.[65]
The RPE functions like a placenta to ensure that the outer retina is protected from injurious compounds, yet allows the necessary nutrients to
pass into the rod and cone area. RPE cell tight junctions are also part of the
outer blood-retinal barrier, preventing diffusion of even small molecules
into the vitreous humor and ensuring that the metabolites required by the
outer retina can move to *where* they are needed *when* they are needed.[66]

To ensure the required level of needed nutrients pass the RPE barrier,
the basal membrane is highly infolded to produce more surface area. This
role is critical because the rods and cones require a greater blood supply
than any other bodily tissue.[67] The high level of metabolism is due to
the complex chemistry required for vision, this necessitates a higher level
of oxygen and nutrients. The RPE also synthesizes and secretes various

---

[63]Bok and Young, 1994, p. 148.
[64]Bok and Young, 1994.
[65]Martínez-Morales, et al., 2004, p. 766.
[66]Hewitt and Adler, 1994, p. 59.
[67]Hewitt and Adler, 1994, p. 59.

extracellular matrix molecules that must be produced near the location where they are to be used due to their short half-life.

If the photoreceptors were anterior to the neurons, as occurs in the verted design, the blood supply would have to be either directly in the light path of the receptors, or on their side. This arrangement would reduce enormously the number of photoreceptors used for sight. If the pigment epithelium tissue were placed *in front* of the retina, sight would be seriously compromised. The verted design would make vision impossible because the photoreceptors *must* be embedded in the RPE to obtain the level of nutrients required to function.

Does the Backward Design Block Inward Coming Light?

Nerve cell fibers and the small branches of the central retina artery and vein produce minimal hindrance to light reaching the photoreceptors because most cells are 60 to 70 percent water. Consequently, they are largely transparent. When viewed under the microscope, most cells are largely transparent. It is for this reason why stains, such as Eosin-Y and Hematoxylin 2, are required to better visualize the various cell parts. Although myelin, an opaque whitish lipid that coats nerve axons, would block much light, in contrast to most peripheral nerves, nerve fibers in front of the retina are not myelinated. Furthermore, the larger blood vessels and nerve fibers skirt around the *area centralis* where visual acuity is most important.[68] The vertebrate eye is highly effective in spite of the retina reversal because it is a precise visual instrument designed to function with the rods and cones facing away from the light:

> The tissues intervening between the transparent humors of the eye cavity and the optically sensitive layer are microscopically thin. The absorption and scatter of light is ordinarily minor, and functional impairment seldom serious. . . . Red blood cells are poor transmitters of light, but when moving single

---

[68]Gregory, 1976.

file through capillaries cause only a negligible shading of the light sensors.[69]

These facts have forced Dawkins to note that many

> photocells point backwards, away from the light. This is not as silly as it sounds. Since they are very tiny and transparent, it doesn't much matter which way they point: most photons will go straight through and then run the gauntlet of pigment-laden baffles waiting to catch them.[70]

Moving shadows produced by the venules and arterioles are also highly functional because they produce momentary darkness to aid in the rod and cone regeneration. Constant bright light would excessively bleach the photopigment, and the momentary darkness achieved by the existing design aids in their regeneration.

Müller Cells Function as Optical Fibers

Placing the retina neural components in front of the photoreceptors does not produce an optical handicap for yet other reasons.[71] One is that the neural elements are separated by less than a wavelength of light. Consequently, very little or no scattering or diffraction occurs, and the light travels through this area as if it were at near-perfect transparency.

> Radial glial cells called Müller cells in front of the retina have both shape and optical properties that contribute to optimizing light transferal and reducing light scatter.[72] Müller cells "have an extended funnel shape, a higher refractive index than their surrounding tissue and are oriented along the direction of light propagation."[73] The effect provides a "low-scattering

---

[69] Williams, 1992, p. 73.
[70] Dawkins, 1996, p. 170.
[71] Land and Nilsson, 2005, p. 63.
[72] Franze, et al., 2007.
[73] Franze, et al., 2007, p. 8287.

passage for light from the retinal surface to the photoreceptor cells," that functions as fiber optic plates very effective
for low-distortion transfer of light images. Cells thought to
interfere with light transmission are actually highly effective
in *reducing* light scatter and distortion, helping to produce a
sharp image.[74]

Other Functions of the Pigment

The many diverse functions of the retinal pigment in the RPE cells that
are "essential for the normal functioning of the outer retina" include
producing a dark brown to black pigment called melanin.[75] The melanin
functions to absorb most of the light not captured by the retina, preventing
the reflection and scattering of light within the eyeball. This inhibits light
from being reflected off the back of the eye onto the retina, preventing
degradation of the visual image and ensuring that the image cast on the
retina by the cornea and lens remains sharp and clear.

Yet another function of the pigment is to form an opaque screen behind
the optical path of the photoreceptors. This light absorptive property of
the pigment is critical to maintaining high visual acuity. For this reason,
normal retinal function requires that the RPE and photoreceptors be in
close proximity. Lack of the pigment, as in albinism, can cause a variety of
problems such as *fovea hypoplasia*, an abnormal routing of the optic nerve.[76]
As a result of this and other factors, albinism victims lack detailed central
vision.[77]

The Macula

The importance of the RPE is indicated by the fact that one of the most
common causes of blindness in the developed world, macular degenera-

---

[74]Franze, et al., 2007.
[75]Hewitt and Adler, 1994, p. 67.
[76]Oetting and King, 1999; Lyle, et al., 1997; Jeffery and Williams, 1994.
[77]Snell and Lemp, 1989; Williamson, 2005.

tion, is the result of RPE deterioration.[78] In this disease, the eye's macula loses its ability to function, causing major central vision loss. Without the nourishment and waste removal role of the pigment epithelium, retina cells will die. Among the other diseases affecting the macula is *central serous retinopathy*, an ion pump malfunction and/or a result of choroidal vascular hyperpermeability.

Detached Retina and the Role of Pigment Epithelial Cells

> The retina is connected to the RPE largely by the interphotoreceptor matrix. When the retina pulls away from the RPE at the interphotoreceptor matrix area, a *detached retina* results.[79] The RPE can then no longer effectively function to regenerate the rods and cones, causing vision to become distorted, and, eventually, the death of a significant level of retina tissue. Progressive detachment can often be halted by laser therapy, a procedure that is only minimally invasive because laser light is able to pass through the cornea and the lens without damaging them. Laser therapy stimulates the migration of the RPE cells, inducing the pigmentation line to form.

The Retina Pigmented Epithelium's Role in Development

RPE is also critical for normal vertebrate eye development. A series of reciprocal cellular interactions during vertebrate eye development determine the fate of the eye components, and the "presence of the RPE is required for the normal development of the eye *in vivo*. Its presence early in development is necessary for the correct morphogenesis of the neural retina."[80] The RPE plays a succession of other roles during embryonic development, including trophic influence, transport functions, retinomotor

---

[78] Zhang, et al., 1995.
[79] Zamir, 1997.
[80] Raymond and Jackson, 1995, p. 1286.

response, and phagocytic and inductive interaction.[81]

Other Possible Designs

A major concern when critiquing the existing vertebrate retina design involves speculation on the quality of vision that would result from another design. No evidence exists that a verted human retina design, as exists in octopi, would result in better vision, and the evidence is clear that it would be worse. Comparisons of different eyes are difficult to make because, although the quality of the image projected on the retina can be evaluated by a study of the lens system's optical traits, direct knowledge about the actual image produced in the brain is lacking.

If the retina were reversed, the retinal pigment epithelium, or its analog, and its cellular support system would have to be placed either in front of the photoreceptors or on their side. Both of these approaches are clearly inferior to the existing vertebrate system that produces superior sight for terrestrial animals. If located in front of the retina, depending on the transparency of these cells, this design would prevent most incoming light from reaching the photoreceptors.

If the RPE functioning cells were located on each side of the rods and cones, as in the cephalopods, only the sensory cell face would be able to respond to light. Octopi use support cells located next to the light sensitive cells called rhabdomeric receptors that use photopigments containing microvilli.[82] The support cells also require increasing the space between the photoreceptors, further decreasing light able to strike the photoreceptors, and consequently lowering vision resolution. The cephalopod's side design "is protective and shields the receptors from excess light."[83] Opaque wastes would accumulate in the light path, and the presence of required nutrients would further diminish the amount of light reaching the photoreceptors. Recycling the outer segments to allow rapid regeneration of the photoreceptors would also be a major problem

---

[81]Coulombre, 1994.
[82]Land and Nilsson, 2005, p. 64.
[83]Prince, 1956, p. 343.

if the photoreceptors faced the vision light path line. Verted designs also produce the following concerns:

> Should the disk end of the rods and cones be reversed in direction so as to face the light . . . we would probably have a visual disaster. What would perform the essential function of absorbing some 10,000 million disks produced each day in each of our eyes? They would probably accumulate in the vitreous humor region and soon interfere with light en route to the retina. If the pigment epithelium layer were placed on the inside of the retina so as to absorb the disks, it would also interfere with light trying to reach the rods and cones.[84]

Although higher visual acuity may improve night vision, in humans it would result in difficulty seeing during daylight hours, which would not be functional for persons that must work in normal human-light environments.[85] Actually, a case can be made that *more* light blockage to reduce light reaching the retina would be functional for many people. Many blue- eyed persons must wear sunglasses for the reason that normal outdoor light is often too bright. In a review of the literature, Young found that excess solar radiation can be a serious health problem, and may

> explain the distinctive global pattern of age-related cataract among human populations—the risk of cataract depends on where one lives on the surface of the earth. . . . Current evidence provides the basis for the design of protective lenses that minimize the hazards of sunlight exposure without sig-nificantly interfering with vision. The prescription has two components—one to protect the lens, the other to protect the retina. . . . Use of sunglasses . . . should begin early in child-hood and be continued throughout the life span whenever exposure to bright sunlight is desirable or necessary. Radia-tion damage to delicate ocular structures can occur at any age and tends to be cumulative.[86]

---

[84]Roth, 1998, p. 109.
[85]Sjostrand, 1989.
[86]Young, 1992, pp. 335-357.

Albinos lack iris pigment, requiring them to wear sunglasses in daylight because even moderately bright light may severely adversely affect their vision.[87] Even blue-eyed persons are at a disadvantage because the lack of brown pigment allows in more light to reach the retina. Consequently, they suffer from more vision problems.[88] The more pigment, the greater the protection. Being able to effectively read by very dim light may be an improvement in some situations, but since most human activities occur during daylight hours, and darkness is functional to induce sleep due to pineal gland activity, the existing system appears to be the most effective.

Furthermore, although the light yellow tint of the eye lens filters out some eye damaging ultraviolet light, the inverted eye design serves to filter out much of the remaining ultraviolet light. The incoming light must pass through the overlying neural components and blood vessels, and the penetrating power of ultraviolet light is markedly inferior to white light.[89] The verted eye is used in animals that live underwater, such as the octopus, where most of the ultraviolet light is filtered out by the water. Consequently, they have less need for this protection. Given the role of the pigmented iris, it is clear that the existing design is ideal.

Conclusions

A review of research on the vertebrate retina consistently supports the view that the vertebrate eye design is perfectly suited for the environment the organism normally lives in. Likewise, the system used by the most advanced cephalopods is also well designed for its environment.[90] The design was intentionally maximized for life in our environment, and would no doubt function poorly in another environment, such as that experienced by undersea bottom dwellers. The RPE metabolic machinery is "essential for the normal functioning of the outer retina [and] ... it

---

[87]Tortora and Grabowski, 1996, p. 461.
[88]Young, 1992.
[89]Lumsden, 1994.
[90]Bergman, 2000; Bergman and Calkins, 2005; Wieland, 1996; Marshall, 1996.

is essential that the RPE and photoreceptors be in close proximity" for normal retina function.[91] This review supports Hamilton's conclusion that

> instead of being a great disadvantage, or a "curse" or being incorrectly constructed, the inverted retina is a tremendous advance in function and design compared with the simple and less complicated verted arrangement. One problem amongst many, for evolutionists, is to explain how this abrupt major retinal transformation from the verted type in invertebrates to the inverted vertebrate model came about as nothing in paleontology offers any support.[92]

Rather than being fired, our camera designer would no doubt be promoted for utilizing a less obvious, but far more functional design. It is clear that "eyesight is a compelling testimony to creative design."[93] This short review covers only a few of the many reasons for the superiority of the existing mammalian retina design. Gratitude rather than impertinence seems the more appropriate response to its ingenious design.

---

[91]Hewitt and Adler, 1994, p. 67.
[92]Hamilton, 1985, p. 63.
[93]DeYoung, 2002, p. 190.

References

Abbott, Joan et al. 1995. *Cephalopod Neurobiology*. New York: Oxford University Press.

Ayoub, George. 1996. "On the Design of the Vertebrate Retina." *Origins and Design* 17 (1): 19-22. Winter.

Barash, David P. and Ilona A. Barash. 2000. *The Mammal in the Mirror: Understanding Our Place in the Natural World*. New York, NY: W.H. Freeman.

Barnes, Robert D. 1980. *Invertebrate Zoology*. Philadelphia, PA: Saunders.

Baylor, D. A.; T.D. Lamb and K.W. Yau 1979. "Response of Retinal Rods to Single Photons." *Journal of Physiology*. 288: 613-634.

Benson, Eliot. 1996. "Retinitis Pigmentosa: Unfolding its Mystery." *Proceedings of the National Academy of Science USA* 93/4526-4528. May.

Bergman, Jerry. "Is the Inverted Human Eye a Poor Design?" *Journal of the American Scientific Affiliation*. 52(1):18-30, March 2000.

_____and Joseph Calkins. "Is the Backwards Human Retina Evidence of Poor Design?" *Impact*. 388:1-4, October, 2005.

Bok, Dean and Richard Young. 1994. "Phagocytic Properties of the Retinal Pigment." In Zinn and Marmor.

Bok, Dean. 1994. "Retinal Photoreceptor disc shedding and pigment epithelium phagocytosis." In Zinn and Marmor pp. 81-94.

Bridges, C.D.B. 1989. "Distribution of Retinol Isomerase in Vertebrate eyes and its Emergence During Retinal Development." *Vision Research*. 29 (12): 1711-1717.

Budelmann, B.V. 1994. "Cephalopod Sense Organs, Nerves and the Brain: Adaptations for high performance and life style." In *Physiology of Cephalopod Mollusks*. Ed by Hans Portner et al. Australia Gordon and Breach pp. 13-33.

Coulombre, Alfred. 1994. "Roles of the Retinal Pigment Epithelium in the Development of Ocular Tissue" (Chapter 4 in Zinn and Marmor).

Dalton, Rex. 2004. True Colours" *Nature*. 428:596-597.

Dawkins, Richard. 1986. *The Blind Watchmaker*. New York: W. W. Norton.

_____. 1996. *Climbing Mount Improbable*. New York: W.W. Norton.

Dennett, Daniel. 2005. "Show Me the Science." *The New York Times*, August 28.

Diamond J. 1985. "Voyage of the Overloaded Ark." *Discover*. June pp. 82-92.

Dowling, John E. 1987. *The Retina: An Approachable Part of the Brain*. Cambridge, MA: The Belknap Press of Harvard University Press.

Edinger, Steve. 1997. "Is there a Scientific Basis for Creationism?" *The Congressional Quarterly Researcher*. 7 (32): 761. August 22.

Ferl, Robert, and Robert A. Wallace. 1996. *Biology, The Realm of Life*. New York: Harper Collins.

Franze, Kristian, Jens Grosche, Serguei N. Skatchkov, Stefan Schinkinger, Christian Foja, Detlev Schild, Ortrud Uckermann, Kort Travis, Andreas Reichenbach, and Jochen Guck. 2007. "Müller Cells are Living Optical Fibers in the Vertebrate Retina." *PNAS*, 104(20):8287-8292.

Frymire, Philip. 2000. Impeaching Mere Creationism. San Jose, CA. Writers Club Press.

Gregory, Richard. 1976. *Eye and Brain*. NY: World Universal Library.

Grzimek, Bernhard. 1972. *Grzimek's Animal Life Encyclopedia*. New York. Van Nostrand Reinhold.

Hamilton, H. S. 1985. The Retina of the Eye--An Evolutionary Road Block. *CRSQ* 22:59-64 Sept.

Hewitt, A.T. and Rubin Adler. 1994. *The Retinal Pigment Epithelium and Interphotoreceptor Matrix: Structure and Specialized Functions*. In Ryan Ed *The Retina* 2nd Ed. St Louis: Mosby pp. 58-71.

Jeffery, G. and A. Williams. 1994. "Is abnormal retinal development in albinism only a mammalian problem? Normality of a hypopigmented avian retina." *Experimental Brain Research*. 100(1):47-57.

Kolb, Helga. 2003. "How the Retina Works." *American Scientist*, 91:28-35.

Kröger, Ronald H.H. Oliver Biehlmaier. 2009. Space Saving Advantage of an Inverted Retina. *Vision Research*. 49:2318-2321.

Kuwabara, Toichiro. 1994. "Species Differences in the Retinal Pigment Epithelium." Chapter 5 in Zinn and Marmor.

Land, Michael F. and Dan-Eric Nilsson. 2005. *Animal Eyes*. Oxford, NY: Oxford University Press.

Lumsden, Richard.1994. "Not So Blind a Watchmaker" *CRSQ* 31:13-21.

Lyle, William M, Jeff O.S. Sangster, and T. David Williams.1997. "Albinism: An Update and Review of the Literature." *Journal of the American Optometric Association*, 68(10)623-645.

Marshall, George. 1996. "An Eye for Creation: An Interview with Eye-Disease Researcher Dr. George Marshall, University of Glasgow, Scotland." *Creation*, 18(4):19-21, September.

Martínez-Morales, Juan Ramón; Isabel Rodrigo, and Paola Bovolenta. 2004. "Eye Development: A View from the Retina Pigmented Epithelium." *BioEssays*, 26:766-777.

Meglitsch, Paul. 1972. *Invertebrate Zoology*. New York: Oxford.

Miller, Kenneth R. 1994. "Life's Grand Design" *Technology Review*. 97 (1): 25-32.

_____. 1999. *Finding Darwin's God: A Scientist's Search for Common Ground between God and Evolution*. New York: Cliff Street Books.

Oetting, W.S. and R.A. King. 1999. "Molecular basis of albinism: Mutations and polymorphisms of pigmentation genes associated with albinism." *Human Mutatagens*. 13(2):99-115.

Pechenik, Jan. 1991. *Biology of the Invertebrates*. Dubuque, IA: Wm. C. Brown.

Peet, J. H. John. 1999. "Creation in the News; Dawkins Blind Spot." *Origins*. 26:2-4 June.

Prince, Jack. 1956. *Comparative Anatomy of the Eye*. Springfield: Charles Thomas.

Raymond, Sophie M. and Ian J. Jackson.1995. "The Retinal Pigment Epithelium is Required for Development and Maintenance of the Mouse Neural Retina." *Current Biology*. 5: 1286- 1295.

Roth, Ariel. 1998. *Origins*. Hagerstown, MD: Review and Herald.

Ryan, Steven J. (Editor).1994. *The Retina*. Second Edition. St Louis:

Mosby.

Sarfati, Jonathan. 1998. A review of "Climbing Mount Improbable" by Richard Dawkins. *Cen Tech J.* 12 (1) 29-34.

Shermer, Michael. 2005. *Science Friction: Where the Known Meets the Unknown.* New York: Henry Holt/Times Books.

Shier, David, Jackie Butler and Ricki Lewis. 1999. *Hole's Human Anatomy and Physiology.* Dubuque, IA: Wm. C. Brown.

Sjostrand, Fritiof. 1989. "An Elementary Information Processing Component in the Circuitry of the Retina Generating the On-Response." *Journal of Ultrastructure and Molecular Structure Research* 102: 24-38.

Snell, Richard and Michael Lemp. 1989. *Clinical Anatomy of The Eye.* Boston: Blackwell Scientific Pub.

Spalton, David J, Roger A. Hitchings, and Paul A. Hunter. 1994. *Atlas of Clinical Ophthalmology.* 2nd ed. Chapter 12: "Vitreou and Vitreo-retinal Disorders." St Louis: Mosby.

Spigel, Irwin M. (Editor). 1965. *Readings in the Study of Visually Perceived Movement.* New York: Harper & Row.

Steinberg, Roy H., and Irmgard Wood. 1994. "The Relationship of the Retinal Pigment Epithelium to the Photoreceptor Outer Segment in the Human Retina." (Chapter 2 in Zinn and Marmor)

Thwaites, William.1982. "Design, Can We See the Hand of Evolution in the Things it has Wrought?" *Proceedings of the 63rd Annual Meeting of the Pacific Division; American Association of the Advancement of Science.* 1 (3): 206-213.

Tortora, Gerald and Sandra Grabowski. 1996. *Principles of Anatomy and Physiology.* New York: Harper and Collins.

Wells, Martin John. 1978. *Octopus; Physiology and Behavior of an Advanced Invertebrate.* London: Chapman and Hall.

Wieland, Carl. 1996. "Seeing Back to Front: Are Evolutionists Right When They Say Our Eyes are Wired the Wrong Way?" *Creation,* 18(2):38-40, March.

Williams, George C.1992. *Natural Selection: Domains, Levels, and Challenges.* New York. Oxford University Press.

————. 1997. *The Pony Fish Glow and other Clues to Plan and Purpose in Nature.* New York: Basic Books.

Young, J. Z. 1971. "The Anatomy of the Nervous System." *Octopus Vulgaris.* New York: Oxford University Press.

Young, Richard. 1992. "Sunlight and Age-Related Eye Disease." *Journal of the National Medical Association.* 84 (4): 353-358.

Zamir, Ehud. 1997. "Central Serous Retinopathy Associated with Advenocorticotrophic Hormone Therapy." *Graefes Archives for Clinical Ophthalmology.* 235: 339-344.

Zhang, Kang, E. Nguyen, A. Crandall and L. Donoso.1995. "Genetic and Molecular Studies of Macular Dystrophies: Recent Developments." *Survey of Ophthalmology.* 40 (1): 51-61.

Zinn, Keith M. and Marmor, Michael (Editors) F. 1994. *The Retinal Pigment Epithelium.* Cambridge; MA: Harvard University Press.

Chapter 11

The Retina Blind Spot

It is often claimed that the human retina is poorly designed because it appears to be placed in the eye backwards. This design requires that light travel *through* the nerves and blood vessels in order to reach the photoreceptor cells which are located *behind* the eye's wiring. This design also requires the nerves to pass through the retina in an area that lacks photoreceptors, producing a blind spot.

One of several reasons why the blind spot is not a problem is because the brain uses information from the retina only to construct an image. It does an excellent job of dealing with the many "blind spots" such as floaters, shadows, reflection problems, dim light, and dust on a person's glasses. Another reason is the area in which the blind spot is located is used only for peripheral vision to scan areas outside what the person is focusing on for areas of potential interest.

A major argument for the existence of a creator is the "argument from design," a conclusion that means the design existing in creation proves the existence of an intelligent designer. Darwinists attempt to disprove the argument from design by providing examples of what they claim display poor design to argue that the living world was not designed but is the result of blind, natural impersonal forces. This view is called the *blind watchmaker* thesis by the former Oxford professor and atheist Richard Dawkins.[1] The human retina is claimed to be one of the most common

[1]Dawkins, 1986.

169

Figure 11.1: Illustration Showing the Blind Spot

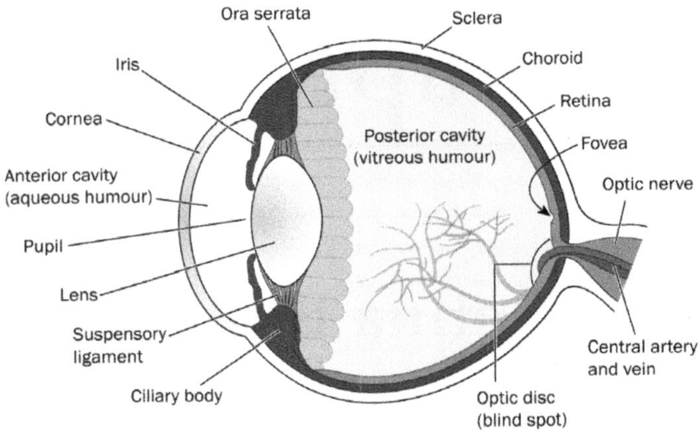

Image Credit: Blamb / Shutterstock.com

examples of poor design in both the popular and scientific literature. The claim is that the vertebrate eye is functionally suboptimal because the retina photoreceptors are oriented *away* from incoming light requiring a blind spot to function.[2] This is alleged to be an example of poor design because, Dawkins reasons, an

> engineer would naturally assume that the photocells would point towards the light, with their wires leading backwards towards the brain. He would laugh at any suggestion that the photocells might point away from the light, with their wires departing on the side *nearest* the light. ..... The wire has to travel over the surface of the retina, to a point where it dives *through* a hole in the retina (the so-called 'blind spot') to join the optic nerve.[3]

Dawkins claims that the problem is "light, instead of being granted an unrestricted passage to the photocells, has to pass through a forest of

---

[2]Bergman, 2000; Ayoub, 1996, p. 19.
[3]Dawkins, 1986, p. 93.

connecting wires, presumably suffering at least some attenuation and distortion," but Dawkins admits the distortion is "probably not much" but "it is the principle of the thing that would offend any tidy-minded engineer!"[4] Tuff's University professor Daniel Dennett argued that, although the human eye design is brilliant, the

> nerve fibers that carry the signals from the eye's rods and cones (which sense light and color) lie on top of them, and have to plunge through a large hole in the retina to get to the brain, creating the blind spot. No intelligent designer would put such a clumsy arrangement in a camcorder, and this is just one of hundreds of accidents frozen in evolutionary history that confirm the mindlessness of the historical process.[5]

Williams claimed the retina is not just an example, but one of the *best examples*, of "poor design" in vertebrates that proves the "blind watchmaker" called natural selection, created life and not an intelligent creator:

> Every organism shows features that are functionally arbitrary or even maladaptive. . . . *My chosen classic is the vertebrate eye.* ... Unfortunately for Paley's argument, the retina is upside down. The rods and cones are the bottom layer, and light reaches them only after passing through the nerves and blood vessels.[6]

Williams admits that the vertebrate eye still functions very well in spite of the backward retina design, but argues that this fact does not negate the backward retina argument because "maladaptive design, however minimal in effect, spoils Paley's argument that the eye shows intelligent prior planning, and the visual effect is real and routinely demonstrable."[7] Barash and Barash even claim that the human

> eye, for all its effectiveness, has a major design flaw. The optic nerve, after accumulating information from our rods and

---

[4]Dawkins, 1986, p. 93.
[5]Dennett, 2005.
[6]Williams, emphasis mine, 1992, p. 72.
[7]Williams, 1992, p. 73.

cones, does not travel directly inward from the retina toward the brain as any minimally competent engineer would demand. Rather, for a variety of reasons related to the accidents of evolutionary history ... optic-nerve fibers first head away from the brain, into the eye cavity, before coalescing and finally turning 180 degrees, exiting at last through a hole in the retina and going to the brain's optic regions. The *result is a small blind spot in each eye* where the optic nerve leaves the cavity of the eye itself, where it should never have strayed in the first place.[8]

He adds: "By contrast, the eyes of an octopus, lacking the troublesome historical constraints of human eye evolution, gather appropriately on the far side of the retina, from which they travel directly to the brain. There is no blind spot."[9] Ayala, under the heading "In Praise of Imperfection," admits the blind spot is only a "minor imperfection, but still an imperfection," which is a difficulty in "attributing the design of organisms to the Creator." The problem, he claims, is that

imperfections and defects pervade the living world. Consider the human eye. The visual nerve fibers in the eye converge to form the optic nerve, which crosses the retina (in order to reach the brain) and thus creates a blind spot, a minor imperfection, but an imperfection of design, nevertheless; squids and octopuses do not have this defect. Did the Designer have greater love for squids than for humans and, thus, exhibit greater care in designing their eyes than ours? [10]

Professor Resnick opines that the human eye is a

perfect example of incompetent engineering. There is a place on the retina where all the nerves go out of the eyeball to convey the information from the receptors to the brain. There

---

[8] Barash and Barash, 2000, p. 296, emphasis mine.
[9] Barash and Barash, 2000, p. 296.
[10] Ayala, 2007, p. 154.

are no rods or cones in that area. This place is called the blind spot and all human beings have it. The blind spot is just bad engineering.... Intelligent design? Not hardly. Any decent optical engineer could have designed a better eye.[11]

Frymire writes that the backward retina is a "classic" example of the many "stupid features which support the idea that they are the result of evolution by natural selection."[12] This topic is also of no small interest to creationists. As Diamond noted, of all of our features

none is more often cited by creationists in their attempts to refute natural selection than the human eye. In their opinion, so complex and perfect an organ could only have been created by design. Yet while it's true that our eyes serve us well, we would see even better if they weren't flawed by some bad design.[13]

A leading American Darwinist, Brown University professor Kenneth Miller, also claimed that a *prime example* of "poor design" is the human retina, a design that Miller claims reflects poorly on an intelligent designer and to Miller provides clear evidence that no designer God exists. Rather, it demonstrates to him that the eye was not designed but, instead, evolved by mutations selected by natural selection.

Miller argues an intelligent designer would not have placed the "neural wiring of the retina on the side facing incoming light" because this arrangement "produces a blind spot at the point that the wiring is pulled through the light-sensitive retina to produce the optic nerve that carries visual messages to the brain."[14] This conclusion is based not only on the incorrect assumption that placing nerves and blood vessels in *front* of the retina reduces the retina's overall effectiveness but, because it creates a blind-spot, another design would be superior. Thwaites argues that the inverted retina problem hits at the *core* of the design argument, and the design argument was historically a major basis for theism:

---

[11]Resnick, 2007, p. 1.
[12]Frymire, 2000, p. 36.
[13]Diamond, 1985, p. 91.
[14]Miller, 1999, p. 10.

Another example straight out of creationist tracts involves the vertebrate eye that humans must share with the other vertebrates . . . the vertebrate eye shows poor design when compared to the eye evolved by the cephalopods. The vertebrate eye has a blind spot where the retinal nerves and the blood vessels exit the eye. There is no comparable blind spot in the cephalopod eye. The structures of the retinas spell the difference.... Clearly the cephalopod solution to retinal structure is more logical, for they have the photosensitive elements of the retina facing the light. Certainly the creationists need to explain why we got the inferior design. I had thought that people were supposed to be the Creator's chosen organism.[15]

And, last, the blind spot is given as a major evidence of the "from the goo to you by the way of the zoo" evolution:

It is because we evolved from sightless bacteria, now found to share our DNA, that we are so myopic. These are the same ill-designed optics, complete with deliberately "designed" retinal blind spot, through which earlier humans claimed to have "seen" miracles "with their own eyes." The problem in those cases was located elsewhere in the cortex, but we must never forget Charles Darwin's injunction that even the most highly evolved of us will continue to carry "the indelible stamp of their lowly origin."[16]

The so-called backward retina is considered a suboptimal design primarily due to its simplistic comparison with a camera. In Diamond's words, requiring light to pass through the nerves and blood vessels is the opposite of how an engineer would have designed the eye and "a camera designer who committed such a blunder would be fired immediately."[17] And Edinger concluded "the vertebrate eye is like a camera with the film loaded backward. . . if an engineer at Nikon designed a camera like that,

---

[15]Thwaites, 1982, p. 210.
[16]Hitchens, 2007, p. 82.
[17]Diamond, 1985, p. 91.

he would be fired."[18] The camera analogy is very inadequate[19] as has been fully documented elsewhere.[20] Our focus here is only on the blind spot problem.

The Research Findings

Most invertebrates possess a verted eye type, and most vertebrates including mammals, birds, amphibians, and fish, possess an inverted type eye. Most verted eye types are very simple, although a few types such as the cephalopod eye (squids and octopus) are almost as complex as the vertebrate eye.[21] Even the better verted eyes are still "overall quite inferior to the vertebrate eye," a conclusion usually determined by measuring performance in response to visual stimuli.[22]

In contrast to Dawkins' claims, no evidence exists that even the most advanced verted eye is superior to the inverted eye. As noted in Chapter 10, the sensitivity of the existing human inverted design is so great that only one photon is able to elicit an electrical response.[23] Consequently, functional sensitivity of the inverted retina could not be significantly improved:

> Neurobiologists have yet to determine how such a negative system of operation might be adaptive, but they marvel over the acute sensitivity possible in rod cells. Apparently rod cells are excellent amplifiers. A single photon (unit of light) can produce a detectable electrical signal in the retina, and the human brain can actually "see" a cluster of five photons—a small point of light, indeed.[24]

---

[18]Edinger, 1997, p. 761.
[19]Werblin and Roska, 2007, p. 73.
[20]Martínez-Morales, *et al.*, 2004; Kolb, 2003; Bergman, 2000; Dowling, 1987.
[21]Land and Nilsson, 2005; Abbott, *et al.*, 1995.
[22]Hamilton, 1985, p. 60.
[23]Baylor, *et. al.*, 1979.
[24]Ferl and Wallace, 1996, p. 611.

Processing Visual Information

Any potential interference of light as it traverses through several layers of the retina before reaching the photoreceptors in an inverted eye is effectively overcome by visual processing. When bipolar or amacrine cells transmit excitatory signals to ganglion cells, the ganglion cells become depolarized, initiating a nerve impulse. Nerve impulses travel along the optic (II) cranial nerve axons which travel to the **optic chiasm** (chiasm means cross) where some fibers cross over to the opposite side of the brain and some remain on the same side.

On the other side of the optic chiasm, the optic tract fibers synapse with neurons in the lateral geniculate nucleus of the thalamus. The lateral geniculate nucleus neurons then form a passageway called the optic radiations to carry the optic information to the primary visual areas in the occipital lobes of the cerebral cortex for extensive processing. It is here where the blind spot is rendered irrelevant.

The Blind Spot (Not *a* Blind Spot)

The blind spot does not reduce vision quality for several reasons. One is that each eye sees a slightly different visual field, and a large area *overlaps.* Although each eye has a blind spot caused by the hole in the retina where the optic nerve (the axons and ganglion cells) passes through in order to travel to the brain, this blind spot falls on a *different* place in each retina.[25] The information from both eyes is then *combined* so that these visual blind spots are not normally perceived.

As a result, because the other eye effectively fills in the gap, special tests are normally required to even notice it. This system not only eliminates flaws, but also produces the binocular visual field that is required to achieve stereo vision. The blind spot is close to 5 in diameter and is located about 15 from the fovea on the temporal side of the visual space.[26]

---

[25]He and Davis, 2001.
[26]He and Davis, 2001, p. 835.

The degree of correction by the brain is so great that it flips the entire image which is upside down on the retina to the right side up view on the occipital lobe of the brain. Light rays from an object in the temporal half of the visual field that faces *away* from the nose will fall in the nasal half of the retina and, conversely, light rays from an object in the nasal half of the visual field will fall on the temporal half of the retina. This reverses the image like when a transparent slide is projected by a slide projector. Also, light rays at the *top* of the visual field strike the *inferior* portion of the retina, and those at the *bottom* portion of the visual field are projected on the *superior* portion of the retina, again reversing the image. Both the left-right and up-down reversal must be corrected by the brain.

Furthermore, filling-in the natural blind spot contributes to binocular rivalry, the necessary slight differences between the two images to form a single three-dimensional image from two slightly different images.[27] The binocular system would be important even if the blind spot did not exist because it enables not only stereovision but also enables the correction system to remove shadows, dirt on one's glasses, eyeball floaters, and other imperfections. Actually, by far the major blind spot is the visual blockage created by the nose, as can easily be seen when one eye is closed. One eye also sees a crescent shaped peripheral monocular visual field that the other eye cannot see, and the same will occur on the opposite side with the opposite eye.

Information received by the brain must be extensively processed in other ways as well. The eye lens is not corrected for chromatic aberration, a fact that all lenses function as prisms separating white light into in colors. The image painted on the retina is "actually rather badly affected by spurious color, and most of the sorting-out [called color correction] is done by the human brain."[28]

This complex operation involves at least three separate systems located in the cerebral cortex, each with a specific function. One system processes information related to *shape*, another regarding *color*, and a third processes information about *movement*, *location* and *spatial organization* of objects. The optical design of the vertebrate eye "approaches

[27]He and Davis, 2001, p. 835.
[28]Watson, 2004, p. 140.

optimal predicted from physics" and in the real world many "animals have a way of confounding the assumptions" based on

> hypothesized models of optimal behavior. In dealing with the interrelated sensory tasks of maximizing spatial acuity and contrast sensitivity, however, both the "camera" eyes of Old World primates and birds, as well as the compound eyes of diurnal insects, present clear examples of evolutionary optimization . . . The investigator's task in examining the hypothesis of optimization is therefore to ask how closely the optical performance of eyes of different optical design approaches the limits set by physics . . . Despite the very different modes of design that underlie the construction of the single-lens eyes of vertebrates and the compound eyes of arthropods, similar considerations determine their capacities to resolve images.[29]

Specialized neurons deep within the retina project what can be thought of as a set of 12 movie tracts, each one a distinct abstraction of the visual world. Each track is sensitive to a representation of only one specific aspect of a scene that the retina must continuously update and stream to the brain. One track may transmit a line-drawing-like image that details only the edges of objects. Another track responds only to motion, often motion in a specific direction. Other tracts carry information about shadows or highlights.[30]

Each track is transmitted by a dedicated set of fibers within the optic nerve to the higher visual centers in the occipital lobe of the brain where even more sophisticated processing occurs. Features such as color, motion, depth, and form are all processed in different regions of the occipital lobe. This fact is known because a lesion in a specific brain region causes a deficit in sensing *one specific* feature.[31] The brain's ability to sense these specific features originates in the retinal "movies."

These "movies" serve as clues that cause the brain to pull its stored "files" on visual images that we all have produced from our life expe-

---

[29]Goldsmith, 1990, pp. 281-282.
[30]Werblin and Roska, 2007, p. 73.
[31]Werblin and Roska, 2007, p. 73.

riences. It is these files that the brain actually "sees." The brain only picks up enough information to pull out the relevant memory file, and occasionally pulls up the wrong file. This is why we occasionally notice someone familiar and, after we speak to him or her, more information input causes us to realize that the person in front of us is not the person that we first thought. Embarrassed, we usually acknowledge our error and walk away.

The Macula

Vision is the sharpest at the macula, an area that is critical in providing the brain with the information needed to construct an image. The macular area is about the diameter of pencil lead but is close to 100 times more sensitive to small visual features than the rest of the retina. It allows us to read, watch TV, recognize friends and even walk. Understanding this structure is another part of the brains' solution to the blind spot problem. Most of the rest of the retina is actually concerned with peripheral vision. The macula provides information required to maximize image detail, and the information obtained from the peripheral areas of the retina where the blind spot is located provides only supplemental information, such as spatial and context details.

An area of the retina in the center of the macula, the **central fovea,** has the highest resolution of the entire retina and, in particular, the foveola. The neurons in this area in front of the photoreceptors are shifted to the side so that light has a *direct* pathway to them, resulting in the least distortion in the area where it matters most.[32] The high-resolution part of the macula also uses more tightly packed cones than the rest of the eye to achieve ultra-high resolution color vision.

The peripheral retina around where the blind spot is located has lower resolution of the retina and consists of mostly rods for black and white vision. The peripheral retina where the blind spot is located functions primarily to survey a large visual area for clues to determine where a person should focus his or her macula for more visual input. Because

---

[32]Goldsmith, 1990, p. 287.

its role is primarily to inform the brain of locations that require more informational input, the peripheral area does not need to pick-up much visual detail. This structure allows the person to be aware of a wide visual field but, at the same time, not be distracted by it.

This design is a highly effective method to accurately transmit enormous amounts of data along the optic nerve in a method analogous to the zipping and unzipping of a computer file to facilitate computer file transmission. To function, the transmission must be very rapid because the image needs to be refreshed continuously like a pixel TV image. The eye's design actually appears to be optimized close to the physical limits of the visible light spectrum.[33] We constantly move our eyes to cause the area of vision sensed by the fovea that is of most interest to us.

If the entire retina was sensitive to the level of detail equal to the macula, the brain would suffer from severe sensory overload. The sensory overload problem is well-documented from research on hyperactivity and auditory sensory overload. If the retina could be reversed so the rods and cones faced in the direction of the light, the peripheral area would require a means of *lowering* the light intensity.

Microsaccades

Microsaccades are tiny unconscious jumps or vibrations that cause the eyeball to move back and forth and from right to left in small jerks. These movements are now known to underpin much of our ability to achieve vision.[34] These movements occur constantly, even when our eyes are fixed on something as they are most of the day.[35] If the movement stopped, eye sight would fade. One reason why is because a static stimulus leads to neural adaptation, causing our neurons to cease responding.[36] Retinal stabilization techniques used in the laboratory causes the visual image to fade away and, conversely, the microsaccades movement prevents the eye

---

[33]Calkins, 1986.
[34]Martinez-Conde and Macknik, 2007, p. 56.
[35]Martinez-Conde and Macknik, 2006, p. 297.
[36]Martinez-Conde, *et. al*, 2004; 2006.

image from fading away.[37] The microsaccades constantly refreshes the image. Diseases that block microsaccades movement also cause vision to fade.

Conclusions

The blind spot, and poor retina design claims in general, are often raised by Neo-Darwinists to argue against intelligent design.[38] Ainsworth and LePage call this the most famous human body design flaw which is "a mistake whichever way you look at" it.[39]

A review of research on the vertebrate retina indicates that the existing inverted design in vertebrates is superior to the verted design, even the system used by the most advanced cephalopods. Its design has been maximized for life in our environment and would no doubt function poorly in another environment, such as that experienced by undersea bottom dwellers. The blind spot does not, even to a minor degree, interfere with vision effectiveness. This review supports Hamilton's conclusion that

> instead of being a great disadvantage, or a "curse" or being incorrectly constructed, the inverted retina is a tremendous advance in function and design compared with the simple and less complicated verted arrangement. One problem amongst many, for evolutionists, is to explain how this abrupt major retinal transformation from the verted type in invertebrates to the inverted vertebrate model came about, as nothing in paleontology offers any support.[40]

As noted in Chapter 10, rather than being fired, our camera designer should be promoted for utilizing a less obvious, but a far more functional, retina design. Gratitude rather than impertinence seems a more appropriate response to its ingenious design.

---

[37] Engbert and Kliegl, 2004.
[38] Peet, 1999; Sarfati, 1998, p. 33; Wieland, 1996.
[39] Ainsworth and LePage, 2007, p. 28.
[40] Hamilton, 1985, p. 63.

References

Abbott, Joan, *et al.*, 1995. *Cephalopod Neurobiology.* New York: Oxford University Press.

Ainsworth, Claire and Michael LePage. 2007. Evolution's Greatest Mistakes. *New Scientist.* Aug 11. pp. 36-39.

Ayala, Francisco J. 2007. *Darwin's Gift: To Science and Religion.* Washington, D.C.: Joseph Henry Press.

Ayoub, George. 1996. "On the Design of the Vertebrate Retina." *Origins and Design* , 17 (1): 19-22. Winter.

Barash, David P. and Ilona A. Barash. 2000. *The Mammal in the Mirror: Understanding Our Place in the Natural World.* New York, NY: W.H. Freeman.

Barnes, Robert D. 1980. *Invertebrate Zoology.* Philadelphia, PA: Saunders.

Baylor, D. A.; T.D. Lamb and K.W. Yau 1979. "Response of Retinal Rods to Single Photons." *Journal of Physiology.* 288: 613-634.

Bergman, Jerry. 2000. "Is the Inverted Human Eye a Poor Design?" *Journal of the American Scientific Affiliation,* 52(1):18-30, March.

Calkins, Joseph. 1986. "Design in the Human Eye." *Bible-Science Newsletter,* March. pp. 1-2.

Dawkins, Richard. 1986. *The Blind Watchmaker.* New York: W. W. Norton.

_____. 1996. *Climbing Mount Improbable.* New York: W.W. Norton.

Dennett, Daniel. 2005. "Show Me the Science." *The New York Times,* August 28.

Diamond J. 1985. "Voyage of the Overloaded Ark." *Discover,* June pp. 82-92.

Dowling, John E. 1987. *The Retina: An Approachable Part of the Brain.* Cambridge, MA: The Belknap Press of Harvard University Press.

Edinger, Steve. 1997. "Is there a Scientific Basis for Creationism?" *The Congressional Quarterly Researcher,* 7 (32): 761. August 22.

Engbert, Ralf and Reinhold Kliegl. 2004. "Microasaccades Keep the Eyes' Balance During Fixation." *Psychological Science,* 15(6):431-436.

Ferl, Robert, and Robert A. Wallace. 1996. *Biology, The Realm of Life*. New York: Harper Collins.

Frymire, Philip. 2000. Impeaching Mere Creationism. San Jose, CA. Writers Club Press.

Goldsmith, Timothy.1990. "Optimization; Constraint, and History in the Evolution of Eyes" *The Quarterly Review of Biology*, 65 (3): 281-322 Sept.

Hamilton, H. S. 1985. "The Retina of the Eye—An Evolutionary Road Block." *CRSQ*, 22:59-64 Sept.

———. 1987. Convergent evolution-Do the Octopus and Human eyes qualify? *CRSQ*,24: 82- 85.

Harris, C. Leon. 1992. *Concepts in Zoology*. New York: Harper Collins.

He, Sheng and Wendy L. Davis. 2001. "Filling-in at the Natural Blind Spot Contributes to Binocular Rivalry." *Vision Research*, 41:835-840.

Hewitt, A.T. and Rubin Adler. 1994. *The Retinal Pigment Epithelium and Interphotoreceptor Matrix: Structure and Specialized Functions.* In Ryan Ed *The Retina* 2nd Ed. St Louis: Mosby p. 58-71.

Hitchens, Christopher. 2007. *God is Not Great: How Religion Poisons Everything*. New York: Twelve.

Kolb, Helga. 2003. "How the Retina Works." *American Scientist*, 91:28-35.

Land, Michael F. and Dan-Eric Nilsson. 2005. *Animal Eyes*. Oxford, NY: Oxford University Press.

Martinez-Conde, Susana, Stephen L. Macknik and David H. Hubel. 2000. "Microsaccadic Eye Movements and Firing of Single Cells in the Striate Cortex of Macaque Monkeys." *Nature/Neuroscience*, 3(3):251-258.

———, Stephen L. Macknik and David H. Hubel. 2004. "The Role of Fixational Eye Movements in Visual Perception." *Nature Reviews/Neuroscience*, 5:229-240.

———, Stephen L. Macknik, Xoana G. Troncoso and Thomas A. Dyar. 2006. "Microsaccades Counteract Visual Fading during Fixation." *Neuron*, 49:297-305.

———and Stephen L. Macknik. 2007. "Windows on the World." *Scientific American*, 297(2):56-63.

Martínez-Morales, Juan Ramón; Isabel Rodrigo, and Paola Bovolenta. 2004. "Eye Development: A View from the Retina Pigmented Epithelium." *BioEssays*, 26:766-777.

Miller, Kenneth R.1994. "Life's Grand Design" *Technology Review*, 97 (1): 25-32.

_____. 1999. *Finding Darwin's God: A Scientist's Search for Common Ground between God and Evolution*. New York: Cliff Street Books.

Peet, J. H. John. 1999. "Creation in the News; Dawkins Blind Spot." *Origins*, 26:2-4 June.

Resnick, Mike. 2007. *Incompetent Design*. http://franksblog. hoferfamily.org/2005/05/05/incompetent-design.

Ryan, Steven J. (Chief Editor). 1994. *The Retina*. Second Edition. St Louis: Mosby.

Sarfati, Jonathan. 1998. A review of "Climbing Mount Improbable" by Richard Dawkins. *T.J.* 12 (1) 29-34.

Thwaites, William.1982. "Design, Can We See the Hand of Evolution in the Things it has Wrought?" *Proceedings of the 63rd Annual Meeting of the Pacific Division; American Association of the Advancement of Science.* 1 (3): 206-213.

Watson, Fred. 2004. *Stargazer: The Life and Times of the Telescope*. Cambridge, MA.: Da Capo Press.

Werblin, Frank and Botond Roska. 2007. "The Movies in Our Eyes: The Retina Processes Information Much More Than Anyone Has Ever Imagined, Sending a Dozen Different Movies to the Brain." *Scientific American*, 296(4):73-79, April.

Wieland, Carl. 1996. "Seeing Back to Front." *Creation*, 18 (2): 38-40 March-April.

Williams, George C. 1992. *Natural Selection: Domains, Levels, and Challenges*, New York. Oxford University Press.

Part IV

Design of the Upper
Respiratory System

Chapter 12

The Maxillary Sinuses

Humans possess four paired *paranasal sinuses*, called *paranasal* because they are located around the nose to supply it with secretions. They are divided into four subgroups named after the bones nearby their locations. The maxillary sinuses, the largest, are under the eyes next to the maxillary bones. The frontal sinuses are located *above* the eyes; the ethmoidal sinuses are *between* the eyes; and the sphenoidal *behind* the eyes.

One often cited common example of poor design is the claim that the maxillary sinus "drain" is upside-down. A comparison is having a kitchen sink drain, not from its base, but from its top. Under the heading "Our Inefficient Sinuses," one evolutionist claimed, "Humans have several sinuses — air-filled cavities that help with drainage of mucus and fluid. But our maxillary sinuses, located on our cheekbone, drain *upwards*. This often leads to the build-up of fluids and mucus, which can cause an infection."[1] In a New York Post article, Getlen opined, "Our horribly designed bodies are making us sick," citing the sinuses as a prime example.[2] Are their claims valid?

The two maxillary sinuses are located above the teeth behind each upper cheek, one on the left side of the nose, the other on the right side. One of several purposes of these four air-filled cavities is to produce a

---

[1]Dvorsky, 2014.
[2]Getlen, 2018.

Figure 12.1: The Maxillary and Surrounding Structures

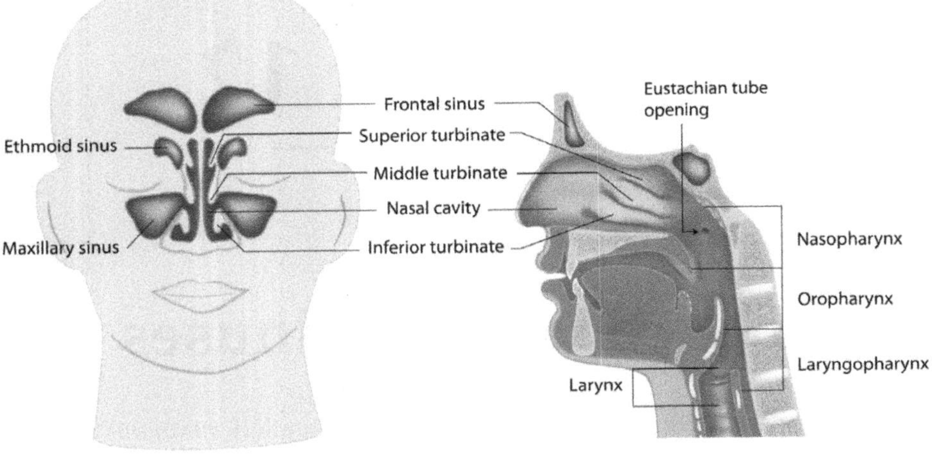

Image Credit: Alila Medical Media / Shutterstock.com

steady supply of mucus and fluids that drain into the nasal passages to moisten and clean the air breathed before it enters the lungs.

The claim is because the mucus discharge duct location at the *top* of the chamber often leads to the build-up of fluids and mucus in the sinus because gravity cannot assist with the mucus movement.[3] Lents claims that, because the maxillary sinuses often must move their contents *against* gravity, they drain poorly, often causing the familiar sinus problems. In fact, pathogens, including viruses and bacteria, cause infections, not the location of the so-called drainage port. In one interview, Lents opines "Few people realize what a mess our nasal sinuses are," adding many

> people are intimately aware of the wretched state of their sinuses, but few are aware of the silly reason why. Our nasal cavities are constantly making mucus to help trap particles, but this mucus must be kept flowing. ..... It turns out that the mucus drainage point of the largest nasal cavity, the one right behind your cheek bones, is placed at the top of the chamber, rather than the bottom. If it weren't for the hard work of the

---

[3]Lents, 2018, p. 11.

hair-like microscopic cilia, constantly propelling the mucus upward, there would no drainage in this cavity ... while we're standing or sitting.[4]

He claims that many other animals are designed with the drainage hole at the *bottom* of the cavity, consequently, he argues, this "is part of the reason why colds and sinus infections are so common in humans but unheard of in other animals."[5] Actually, as any pet lover knows, many animals suffer from sinus infections and related problems just as do humans.[6] Second, for billions of people, their maxillary sinuses normally function without problems for most of their life. And in the vast majority of cases the body can deal with those that may become infected without problems.

Nonetheless, ignoring this fact, Lents claims due to evolution "Humans got the worst of it, by far. The drain pipes of the maxillary sinuses grew skinnier and skinnier [as humans evolved], and, to make a bad situation worse, got stuck at the top of the chamber rather than at the bottom" as is true of other primates, including chimps.[7] There is no evidence that the drainage ports got smaller as humans evolved, and Lentz cites none. African apes, gorillas, and chimpanzees (the currently claimed closest ape to humans), all have four sinuses, as do humans. In contrast, the Asian apes, orangutans and gibbons, the so-called lesser apes, because of their smaller size, possess only two sinuses, the maxillary and sphenoid. They lack both the ethmoid and frontal sinuses.

The poor design advocates often use the term "drain," as if the sole function of the main exit port is to empty the sinuses. Their real purpose is a portal to the nasal chamber steadily supply the nasal cavity with mucus to prevent drying and distribute antibiotics and anti-viruses to protect the lungs against air contaminants and dust. Furthermore, the sinus problem is almost always a mismatch between the production and removal rates of the large amounts of mucus that is generated *only* during infections. This mismatch can occur for several reasons, but usually contributes to

---

[4]Lents, 2018c.
[5]Lents, 2018, p. 11.
[6]Ogilvie and LaRue, 1992; Norris and Laing, 1985.
[7]Lents, 2018c.

the sick symptoms which is the body's way of forcing rest to have the energy required to focus on fighting the infection.

The Poor Design Claimed to be a Result of Evolution

The poor design of the maxillary sinuses is theorized to be a result of evolution.[8] Evolutionary theory postulates that, because the maxillary sinuses are the most frequently infected paranasal sinuses in humans, infection may occur relatively "commonly in the maxillary sinuses due to the position of their ostia [openings into the nasal cavity, (singular ostium, plural ostia)] high on their supermedial [sinus] walls, which may be suboptimal for natural drainage." The claim is that the sinus secretion system is poorly designed due to evolutionary lag, i. e., after humans evolved to walk upright "the ostia remained in a quadrupedal position as bipedal humans evolved from their [quadruped] primate ancestors."[9]

Maxillary sinus drainage is assumed by evolutionists to be optimal in the quadrupedal position, concluding, "human maxillary sinuses exhibit better passive drainage through their ostia when tilted anteriorly to mimic a quadrupedal head position, . . . an example of an evolutionary lag phenomenon, and could be one etiologic factor in the prevalence of maxillary sinusitis in humans."[10]

One study was conducted by filling cadaveric human and goat maxillary sinuses with saline solution in each head position, then measuring the ostia volume drainage. One problem with the study is it should have been completed using a fluid with close to the same viscosity as sinus secretions. Other problems include then fact that cadaveric sinus cavities are often distorted due to age, disease, or distortions after death.

---

[8]Wayman, 2018.
[9]Ford, 2011, p. 70.
[10]Ford, 2011, p. 70.

The Physiology of Sinus Drainage

Sinus drainage involves, not just the visible ostium opening, but a complex system of often microscopic interconnections between the set of paranasal sinuses and the surrounding tissue.[11] Although the primary ostium is located on the *upper* wall of the maxillary sinus, it is not the only, or in some cases, even the main fluid-transport route out of the sinus. In short, the ostia are not designed to be the only means of fluid conduction out of the sinuses.

Fluids from the paranasal sinuses are conveyed via a very complex pathway system, using many accessory ostia and several rather ingenious interdependent pathways. Furthermore, transfer movement is not primarily by "gravity," but by microscopic hair-like cells that line the sinuses causing ciliary action to move secretions into a drainage channel network into the nose.

The Cilia

The major source of movement of the mucus secretions are the cilia. The nose and its accessory cavities' mucous membrane are completely covered with ciliated epithelium. These cilia are in constant motion which has been compared to a whip lashing. All cilia of a "single cell move in the same direction at the same time, but the cilia of all the cells do not move simultaneously."[12]

Rather, the motion of cilia moves over the mucous membrane in a wavelike manner. [13] The general direction the mucus moves is toward the ostia. Furthermore, each cilium moves through an arc of from 20 to 30 degrees at a rate of "about 12 times per second, the forward movement being about twice as rapid as the return movement."[14] This movement

---

[11]Daniels, 2003.
[12]Yankauer, 1908, p. 520.
[13]Yankauer, 1908, p. 520.
[14]Yankauer, 1908, p. 520.

Figure 12.2: Ciliated Epithelium

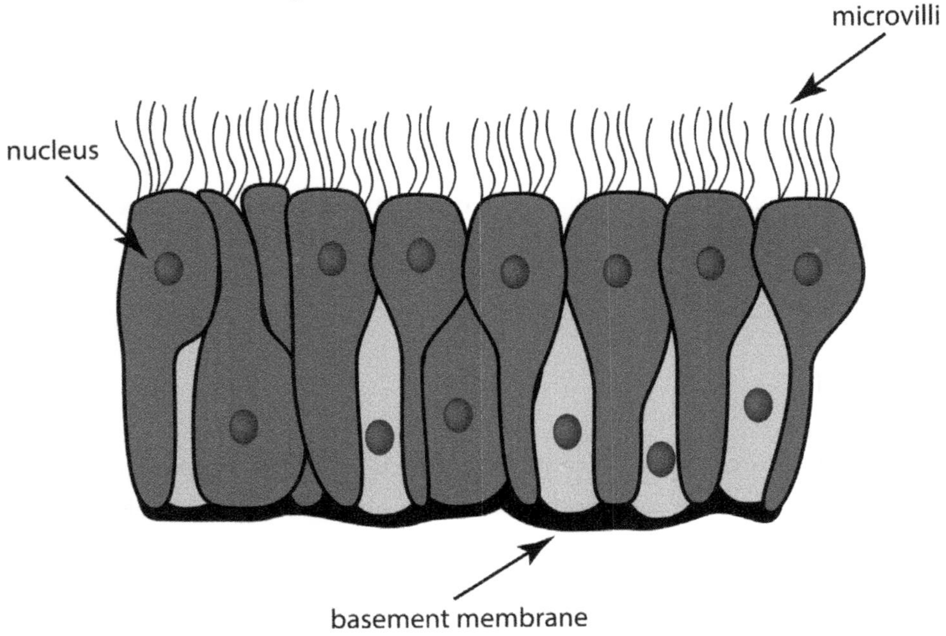

Image Credit: Timonina / Shutterstock.com

occurs constantly, and is the major "factor in maintaining the drainage of these cavities."[15]

As a result, the fact that the sinuses are "capable of draining themselves when they are in a normal, healthy condition will hardly be disputed; but . . . [normally] are also capable of emptying themselves through their natural orifices [even] when they have become diseased."[16] Some persons prone to sinusitis have cilia that don't function properly due to heredity, normal genetic variations, damage due to disease, accident, tumors or polyps.[17]

The cilium move the secretions around the sinus wall cavity primarily to provide moisture and anti-bacteria protection, requiring a steady flow of secretions to achieve this function. It also serves the same function in the nasal passageways. When the sinuses are over-filled, much of the mucus may be removed by the cilia, and forced out through the top, helped by gravity when lying prone. In this case the large ostium on the upper wall of the sinus can function as an "overflow" channel, analogous to the overflow opening in sinks and bath tubs. This overflow system is especially important when significantly higher secretions than normal are produced to help fight infections.

Research on Accessory Ostia

The "Inter-sinus connections and accessory ostia of the maxillary sinus are well-known to rhinologic surgeons but are less known for the remaining paranasal sinuses."[18] A review of "the literature on accessory sinus ostia, inter-sinus connections, and mucociliary drainage pathways for the entire sinus system" has documented accessory

> sinus ostia for each paranasal sinus. Many sinuses drain not only directly into the nasal cavity but also indirectly through adjacent sinuses . . . . one major drainage pathway of the

---

[15] Yankauer, 1908, p. 521.
[16] Yankauer, 1908, p. 518.
[17] Fliegauf, et al., 2007.
[18] Mann, 2011, p. 245.

> frontal sinus is over the ethmoid sinuses and via the ethmoids into the maxillary sinus and subsequently into the nose. . . . Accessory ostia are not only common for the maxillary sinus but also for the entire paranasal sinus system.[19]

The findings of Mann and his colleagues highlight the fact that the poor design claim ignores the complexity of the sinus drainage system and, particularly, the incorrect belief that the large ostium at the top of the maxillary sinus is the only drainage path.

By asking design questions, deeper insight of sinus physiology results which generate new scientific questions relevant to medical practice. The maxillary nasal sinuses cilia move the sinuses' contents into the nose cavity, and the main problem occurs when they become inflamed. Infection can cause similar problems for all of the paranasal sinuses, not just the maxillary nasal sinuses.

Floor Drainage is Problematic

Furthermore, the sinus *floor* drainage ostia design may drain at too high a rate, drying out the sinus mucosa and predisposing plugging the ostium by thick debris. What is required is a *steady flow* of mucus into the nasal cavity. Another reason the ostium is not on the sinus floor is because it would drain into the area *below* the floor of the maxilla, i.e., into the oral cavity (the mouth). This would bypass their purpose, the lubricate and remove dust and dirt from the nose cavity.

The sinuses are designed to drain secretions into the nose also to moisten it with wet and sticky mucus containing anti-bacterial and anti-viral agents to kill bacteria and viruses. Much of the dead pathogen-containing mucus is ultimately swallowed, exposing it to the highly acidic stomach environment that chemically breaks it down, destroying the pathogens. For this reason, the maxillary sinuses' ostia are located in the areas required to serve their central role of coating the nasal passageways

---

[19]Mann, 2011, p. 245.

with mucus. Their location is in an area of the nose where it will best serve this function (see illustration).

Given the nasal sinuses' many functions, including lubricating, warming, and moistening the air we breathe, as well as attacking pathogens, the four different sinuses as a set are designed to completely surround the nasal cavity. All of the paranasal sinuses also function to produce voice resonance to allow each human voice to be unique so that family and friends can rapidly recognize the person talking. When "plugged up" during a cold, the voice is distinctly different, a fact that illustrates this. An advantage of this design is voice analysis can accurately identify persons making phone calls or tape recordings.

Furthermore, "drainage of the accessory sinuses is influenced by gravitation [only] to a slight degree in certain positions of the head."[20] The claim that "poor mucus drainage in the maxillary sinuses, which is a chief reason why humans get the 'common cold' and sinus infections way more than any other animal," as noted above, is demonstrably incorrect.[21]

The common cold and sinusitis usually are caused by a virus and a higher mucus production is part of the bodies response to fight infections. Residue signs and symptoms of the infection often remain even after other upper respiratory symptoms have dissipated. Specially, secretions of inflamed sinuses are much thicker and more viscous than that of healthy sinuses.

The Literature on the Ostia Location

A literature search has failed to document the claim that sinus problems are often due to the incorrect ostia placement claim of the poor design model. Conditions such as allergies, nasal polyps, and tooth infections are major contributors to sinus problems and symptoms. One study noted, "up to 40% of chronic bacterial maxillary sinus infections are attributed to a dental source."[22]

---

[20]Yankauer, 1908, p. 528.
[21]Lents, 2018, p. 11.
[22]Patel and Ferguson, 2012, p. 24.

Furthermore, contrary to popular belief, infections causing inflammation of the sinuses does not always cause facial pain or headaches, but typically symptoms of fullness or pressure. No evidence exists that, instead of infective agents, the ostia location is the cause of sinus problems. Many sinus problems would still exist even if the ostium was located on the sinus floor.[23]

An early dissection study of 114 lateral nasal walls of fresh adult heads, paying special attention to the presence of accessory ostia, was completed by Myerson. He found that the main ostia exist in a variety of shapes and sizes.[24] The size and shape variations are likely a critical reason for drainage effectiveness, or lack thereof. He also located numerous accessory ostia which varied from pinheaded size to several mm in diameter.[25] His literature review found that accessory ostia existed in as many as half of all cases he examined. Problems with this and similar studies, which include the use of cadavers, create the possibility of contamination caused by disease, old age, and factors involved in sample preservation.

Nonetheless, one clear finding was the variation in sizes and shapes of both primary and accessary ostia. This research concluded the critical factor in sinus problems is not the location of the ostia, but very small ostia and/or few or no accessary ostia, a fact that explains why most people never have, or rarely have, sinusitis problems, while others have many problems.

Another study of the maxillary sinus in 37 normal persons showed that the mean functional size of the primary ostium corresponded to a diameter of 2.4 mm, some were smaller, others were larger.[26] This is a relatively large hole, and in these cases, the main problem would develop when bacterial or viral infection causes swelling, partly closing the opening.

A large sample study of several hundred persons concluded that an enormous level of variation of sinus traits existed.[27] This finding supports

---

[23]Gwaltney, 1996; Dykewicz, and Hamilos, 2010.
[24]Myerson, 1932 p. 83.
[25]Myerson, 1932, p. 89.
[26]Aust and Drettner, 1974, p. 432.
[27]Neivert, 1930, p. 1.

the conclusion that sinus problems may, in part, be due to this normal ostia variation. The fact that many of these studies are very old, in some cases close to a century, indicates much has yet to be learned about the maxillary sinus design, and to conclude the problem is poor design is both unwarranted and premature.

Summary

The existing evidence supports the conclusion that the maxillary sinus design is not defective, but rather is optimal for the designed functions of the complex sinus system. In addition, moving the ostium to the maxillary sinus floor would likely be seriously counterproductive. Furthermore, "it is evident that gravitation, as such, plays a very small part in the drainage of the normal accessory sinuses," invalidating Lents claim of poor design based on the location of the drainage portal at the bottom of the sinus.[28]

The sinusitis problem that poor design is blamed for is usually due to one or more of the following conditions: poor health, poor diet, disease, and normal structural variations. This falsifies both the poor design argument and the view that the existing design resulted from an evolutionary lag due to human evolution from quadrupedal primates.

Sinus physiology is very complex, and focusing on the location of the maxillary ostia seriously misrepresents the function of the entire system as a unit. Drainage is mostly by cilia action and fluid exchanges across lymphatic and vascular channels. The "poor design" arguments are generally based on a lack of understanding of their physiology. As summarized by the classic study, still relied on today, the main means of fluid drainage is

> ciliary activity, by the power of the secretion of creeping over the surface of the mucus membrane, i.e., adhesion, by capillary attraction at the natural orifices of the sinuses, and by the syphonage exerted by the nasal passages. The combined

---

[28] Yankauer, 1908, p. 520.

action of these forces is favored by the respiratory air currents, and by certain atmospheric conditions.[29]

The sinuses are amazingly complex and beautifully designed for the many functions they ae designed to accomplish. They work perfectly for 70 plus years for most all of the billions of humans living on Earth today. Each sinus cell is a marvel of design, and each of the millions of organic molecules used to construct them are likewize elegantly engineered.

[29] Yankauer, 1908, p. 528.

References

Aust, R. and B. Drettner, 1974 The Functional Size of the Human Maxillary Ostium in Vivo. *Acta Oto-Laryngological Journal*. 78:432-435.

Daniels, David. 2003. The Frontal Sinus Drainage Pathway and Related Structures. *American Journal of Neuroradiology*. 24(8):1618-1627. September.

Dvorsky, George. 2014.The Most Unfortunate Design Flaws in the Human Body. The Daily Explainer. February 7. https://io9.gizmodo.com/design-flaws-1518242787.

Dykewicz, Mark S. and Daniel L. Hamilos. 2010. Rhinitis and Sinusitis. *Journal of Allergy and Clinical Immunology*, 125(2)S103-115, Supplement 2, February.

Fliegauf, Manfred, Thomas Benzing and Heymut Omran. 2007. When cilia go bad: cilia defects and ciliopathies. *Nature Reviews Molecular Cell Biology* 8:880–893.

Ford R. L, Barsam A; P.Velusami and H. Ellis. 2011. Drainage of the maxillary sinus: a comparative anatomy study in humans and goats. *Journal of Otolaryngology Head and Neck Surgery*. 2011 Feb;40(1):70-4.

Getlen, Larry. 2018 Our horribly designed bodies are making us sick. May 12. https://nypost.com/2018/05/12/our-horribly-designed-bodies-are-making-us-sick/.

Gwaltney, Jack. 1996. Acute Community-Acquired Sinusitis. *Clinical Infectious Diseases*. 23(6): 1209-1223 December.

Lents, Nathan H. 2018. *Human Errors: A panorama of our glitches, from pointless bones to Broken Genes* Houghton Mifflin Harcourt, Boston, MA.

_____. 2018b. The Discovery Institute says they "shellacked" me on 'Human Errors.' Here I defend my claims. https://thehumanevolutionblog.com/2018/05/23/response-creationists-testes-sinuses/.

_____. 2018c The Flawed Human Body: What Would I Change If I Could? http://www.powells.com/post/original-essays/

the-flawed-human-body-what-would-i-change-if-i-could.

Mann, Wilf, J. et al., The Drainage System of the Paranasal Sinuses: A Review with Possible Implications for Balloon Catheter Dilation. *American Journal of Rhinology & Allergy.* 25(4):245-8. July-August.

Myerson, Mervin 1932 The Natural Orifice of the Maxillary Sinus *Achieves of Otolaryngology.* 1932;15(1):80-91.

Neivert, Harry. 1930 Symposium on maxillary sinus: surgical anatomy of the maxillary sinus *The Laryngoscope,* 40(1):1-4.

Norris, Alan and Elizabeth Laing. 1985. Diseases of the Nose and Sinuses. *Veterinary Clinics of North America: Small Animal Practice.* 15(5): 865-890. September.

Ogilvie, Gregory and Susan LaRue . 1992. Canine and Feline Nasal and Paranasal Sinus Tumors. *Veterinary Clinics of North America: Small Animal Practice.* 22(5): 1133-1144, September.

Patel, Nimish and Berrylin Ferguson. 2012. Odontogenic sinusitis: an ancient but under-appreciated cause of maxillary sinusitis. *Current Opinion in Otolaryngology & Head and Neck Surgery.* 20(1):24–28. February.

Wayman, Erin. 2018. Clues to Ape (and Human) Evolution Can Be Seen in Sinuses. https://www.smithsonianmag.com/science-nature/clues-73349254/.

Yankauer, Sidney. 1908. The Drainage Mechanism of the Normal Accessory Sinuses. *Laryngoscope.* 18(7): 518-529.

Chapter 13

The Pharynx

The concept of dysteleology, is the claim that poor design exists everywhere in the natural world, therefore an intelligent creator does not exist. This chapter discusses a common example of dysteleology, the putative poor design of the pharynx, concluding the claims it was poorly designed are misinformed or demonstrably wrong. In fact, it's design shows clear evidence of superior design.

An example is from a noted zoologist teaching at a major university who claimed that the human pharynx is a poorly designed system, explicable only in terms of macroevolution. The example he gives is designing a building with water and gas entering through a common chamber so that whenever one is needed, the other would have to be shut off. He claims this design

> would be the height of stupidity. But that is what your "intelligent Creator" did when he designed and created man for, as you know, the pharynx serves as a common passageway for air and water. Think of the number of lives that have been lost by food or water getting into or obstructing the air passageway. It certainly would have required very little intelligence for the Creator to have designed a more efficient and less dangerous arrangement.[1]

---

[1] letter quoted in Howe, 1981a, p. 3.

Evolution answers the source of this poor design by tracing the hypothetical evolution of the head and also

> the development of the food and respiratory passageways from the fishes up through the amphibians, reptiles and early mammals to man, you will note that the relationship turns out to be a masterpiece of evolutionary achievement enabling aquatic organisms to become adapted to air breathing and thus capable of living on land.[2]

In short, a solution to adaptation to land to enable aquatic animals to adapt to land. University of Michigan Professor evolutionists Scott Atrain writes that the problem is the mouth cavity in air- breathing terrestrial animals

> does double duty, as an opening to take in both food and air. As creatures evolved from water onto land, the opening to the respiratory system was jerry-rigged to share the preexisting digestive tract's anterior structure, including the mouth and pharynx [throat]. In terrestrial vertebrates, the pharynx became a short passage linking the mouth to the esophagus and the windpipe. Any mistiming of the swallowing mechanism that blocks off the air passage in routing food to the esophagus causes choking.[3]

He then claims that the problem for humans is worse because the

> mouth and throat do triple duty, serving also the function of speech. Both in swallowing food and in articulating speech sounds, respiration is temporarily inhibited, as the larynx rises to close (in swallowing) or constrict (in speaking) the opening to the air passage (glottis).[4]

He next claims, without citing any evidence, that humans are more likely

---

[2]letter quoted in Howe, 1981a, p. 3.
[3]Atrain, 2006, pp. 129-130.
[4]Atrain, 2006, pp. 129-130.

than other animals to choke when they attempt to coordinate eating, breathing, and speaking. In the bargain, the swallowing ability of humans has become much weaker than that of other animals. So when moms tell their kids, "Don't talk while you're eating!" they're helping to make up for an evolutionary failing.[5]

Ardea Skybreak, in her book endorsed by Richard Leakey and University of California, Berkeley Professor Kevin Padian, wrote that humans "have a dangerous tendency to choke on food" because the

passage that air follows to get to the lungs actually *crosses* the path that food allows to get to the stomach. This would be an example of a really stupid (or perversely sadistic) design *if* a god had actually designed it that way! But this is not the result of anyone's conscious "design." This choking problem simply reflects our own past evolutionary history: the breathing channels of all land vertebrates also evolved in the distant past as *modifications of pre-existing structures* (in this case the "air bladders" of some bony fishes and lungfishes) which evolved into primitive lungs.[6]

She concluded that the air and food passages crisscrossing

"innovation" allowed the first land vertebrates to breathe out of water and to spread into many new habitats. But as much as it created all sorts of new opportunities and advantages, allowing new species to spread and diversify across the land, this evolutionary development came "packaged" with this little problem of air and food passages "crossing."[7]

Professor Carey Carpenter wrote that he often cites

examples of "incompetent design" to my anatomy students. I do this to counter the neo-creationist notion of "Intelligent De-

---

[5] Atrain, 2006, pp. 129-130.
[6] Skybreak, 2006, p. 109; emphasis mine.
[7] Skybreak, 2006, p. 109; emphasis mine.

sign" that is creeping into some curricula. For example, I discuss how the anatomy of the pharynx sets the stage for thousands of choking deaths every year, and how, in my opinion, a common pathway for both food and air would simply not be something a competent, intelligent designer would come up with! I follow up with a discussion of how evolutionary theory can account for such a jury-rigged structure. I have other examples as well (the blind spot in the eye for one).[8]

An analysis of the pharynx's design eloquently falsifies the poor design claim. In fact, the pharynx is an example of a superbly designed complex system. The pharynx serves as a single passage for three different systems—the respiratory, digestive, and communicative—for many very good reasons. A major one is, unlike other primates, our airway and esophagus intersect. This can cause choking, but it also is what allows speech.[9]

The pharynx connects the air channel to the alimentary canal. This design allows disposal of both excess moisture in the air channel, and the dust and other debris in the lung system that is filtered from the air by bronchial mucus. The mucus acts like fly paper, sticking to dust when it is moved up out of the lungs by cilia and is then swallowed where the stomach digests and disposes it. This design also allows the creation of air pressure bursts, a response called coughing or sneezing that are necessary to remove irritants from the throat and nose. This system is critical to force out objects, such as dust and other irritants as well as food that can occasionally get stuck in the area of the food tube above the epiglottis or in the back of the mouth.

The pharynx structure permits the mouth and nose to alternate as breathing ports—a feature that is critical whenever the nose is plugged, as when suffering from a cold, or the mouth is blocked, such as when it contains food. The nostrils are used when there is a need for breathing normal quantities of warmed, humidified, filtered air, and the oral cavity allows rapid entry of much larger quantities of air when needed.

---

[8]Carpenter, 1999.
[9]Walter, 2006.

Figure 13.1: The Pharynx and the Surrounding Structures

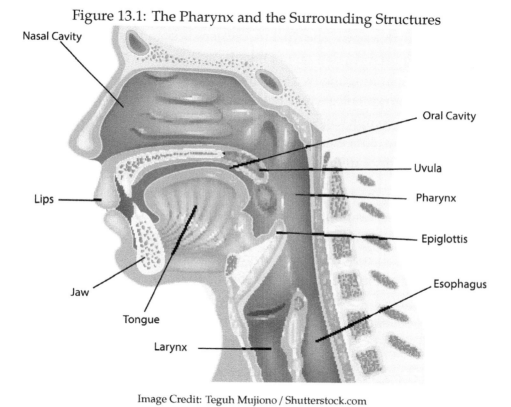

Image Credit: Teguh Mujiono / Shutterstock.com

The tongue, teeth, palate, mandible, and cheeks are all necessary for manipulating food, but they are also all required for speech.[10] These structures, called articulators, have critical functions in the formation of vowel sounds.[11] The two-tube design requires two separate mouths, lips, tongues, teeth and other structures, duplicating many of the same structures, using one set for eating and the other set for speaking.

The pharynx design allows both simultaneous eating and breathing with greater efficiency and less body bulk than if we had two separate unconnected channels. The duel use system is both ingenious and economical. Importantly, one cannot breathe and swallow at the same time, effectively separating the two systems. If deglutition is unimpeded, the food bolus has no choice where to go—it has to enter the esophagus. Critics that argue completely separate tube systems, one for respiration and another for the alimentary tract, is a superior design, have not thought through this very carefully.

This design, though, would require a far more complex tube and networking system, resulting in a greater likelihood for errors and casualties. Another problem with this design is the more body openings, the more difficult it is to protect the body from pathogens. By using three openings instead of the present one, the likelihood of infections would also increase significantly, and pathogen protection would likewise have to be increased. Given how many pathogens we take into our oral cavity, three oral cavities could cause major problems.

Another problem is that the sense of taste is intimately involved with our sense of smell. The olfactory sense for this reason is used in both eating and digestion, and is also part of the respiratory system that allows us to "taste" our food. Otherwise, food would be tasteless, such as occurs when we have a heavy cold. Separate systems would require a totally different design, which would be impractical.

---

[10]Laitman, 1984.
[11]Gordon-Brannan and Weiss, 2007.

The Human Language System

Humans, unlike apes, have a "descended larynx," meaning it is located much farther down the trachea toward the lungs than it is for all other primates. The larynx in humans actually sinks *lower* as the baby grows until, in adults, it is located at the junction of the food tube and larynx (windpipe) directly below the base of the tongue. The "long larynx" in humans is what allows us to choke, but also allows us to use language very effectively.[12]

Furthermore, so far as is known, this design feature exists in humans only.[13] This design allows speech in humans—the only life form on Earth that has acquired a language—which is a major reason why Darwinists argue that a descended larynx must have evolved first, and only later was speech able to evolve.[14] This design feature also makes gulping large amounts of air very easy, a very useful trait for underwater swimming.

The Abuse and Disease Problem

When food or water enter the wrong tube (the trachea), it is not because the system is poorly designed, but occasionally problems occur because of disease or abuse, such as eating while under the influence of alcohol or someone applying first aid incorrectly. People do not die because of a poorly designed pharynx, but rather because of its injury, abuse or disease. If one eats or drinks too fast, and/or has some pathology associated with the area, then it may not function properly, a fact that is true for the most efficiently designed systems as well.

In support of the poor design claim, its advocates cite statistics on choking. Humans swallow about 1,000 times a day, or 28 million times in an average lifespan. Given this fact,

life-threatening choking is actually a very rare event. It is most com-

---

[12]Rice, 2007, p. 232.

[13]Morgan, 1994.

[14]Morgan, 1994.

mon in very young children, often caused by swallowing small toys, hard candy, or gum—all things that small children should not be allowed to play with.[15] The most common choking problem is with infants under six-months old. Over half are non-food related and most are due to lack of proper parental supervision.

The next largest problem is in the elderly, often caused by disease such as Parkinson's, Alzheimer's, Myasthenia Gravis, Amyotrophic Lateral Sclerosis, Cerebral Palsy, Multiple Sclerosis or strokes or some type of dysphasia. A final group for which choking is a problem is the developmentally handicapped. For healthy adults, the most common problem is eating too fast, trying to swallow very large food portions, talking or laughing while eating—all problems that the Heimlich maneuver can usually solve if properly applied. The Heimlich maneuver is performed by standing behind the person choking and exerting upward pressure on the victims diaphragm to compress the lungs, forcing air out through the trachea to dislodge the obstruction.

The system is so effective that multi-millions of people have consumed three meals a day for a lifetime without problems. Swallowing causes the pharynx to stimulate several very complex reflex responses. The first shuts off the passage into the nose by raising the vellum, and the next one closes the opening of the trachea with a flap called the epiglottis, and another response forces the food down into the top of the esophagus by peristalsis. The human soft palate elevates in order to close off the nasopharynx using a very different design than used in all primates.

One serious problem is the aspiration of food into the lungs can cause pneumonia and can even be lethal. This extremely rare event usually occurs in stroke victims or others who have nervous system damage that interferes with the proper functioning of the complex swallowing system. In the vast majority of cases, problems only arise with a misuse of, or degeneration in, the system. This cannot be used as evidence of bad design because good original design can be negated by unwise or unsafe behavior, as well as the subsequent degeneration of the system.

---

[15]Baker, et al., 1992.

Conclusions

Many critics of the design argument commit the logical fallacy of "special pleading" by calling the human pharynx design "the height of stupidity" when discussing creation, but view the pharynx as a masterpiece of biological engineering when they attribute its design to evolution. This illogical reasoning is unfortunately common in debates on origins. The fact is, the pharynx system serves several functions very efficiently and effectively.

Comparing the oral cavity's design to achieve its many functions to other possible designs, demonstrates the poor design claim is invalid. There are at least a dozen important reasons for its existing design. The only way to scientifically prove that another system is better is to do a scientific comparison of two designs—one that uses the original system, another that uses the improved system. This experiment will never be done, as it would require major surgery and likely would create serious health problems.

Problems with the pharynx occur very rarely if used correctly. In view of the sophistication and anatomical interplay involved in the deglutition reflex—the lips close, the tongue flips up and back, the uvula closes off the nasopharynx, the epiglottis seals off the glottis, all coordinated by the nervous system—it is amazing that problems do not occur more often.

References

Atrain, Scott. 2006. "Unintelligent Design," pp. 126-141 in *Intelligent Thought*. New York: Vantage, John Brockman editor.

Baker, Susan P., Brian O'Neill, Marvin J. Ginsburg, and Guohua Li. 1992. *The Injury Fact Book. Second Edition.* Oxford, NY: Oxford University Press.

Carpenter, Carey. 1999. "Re: Incompetent Design? [HAPP-L]." http://daphne.palomar.edu/ccarpenter/HAPSList/HAPP-L.1999.2/msg00436.html.

Gordon-Brannan, Mary E. and Curtis E. Weiss. 2007. *Clinical Management of Articulatory and Phonologic Disorders.* Philadelphia, PA: Lippincott Williams and Wilkins.

Howe, George. 1981. "Correspondence Series" *Students for Origins Research* 4(1):3, Winter-Spring.

_____. 1982. "Final Remarks on the Pharynx Design. Also: A Challenge for Evolutionists." *Students for Origins Research*, 5(1):10, Winter-Spring.

Laitman, Jeffrey T. 1984. "The Anatomy of Human Speech." *Natural History*, 8:20-27.

Morgan, Elaine. 1994. *The Scars of Evolution: What Our Bodies Tell us about Evolution.* New York: Oxford University Press.

Namm, Ted and UMass Lowell. 1999. "Re: Incompetent Design? [HAPP-L]." 2pp. http://daphne.palomar.edu/ccarpenter/HAPSList/HAPP-L.1999.2/msg00436.html.

Rice, Stanley. 2007. *Encyclopedia of Evolution.* New York: Facts on File.

Skybreak, Ardea. 2006. *The Science of Evolution and The Myth of Creationism: Knowing What's Real and Why It Matters.* Chicago, IL: Insight Press.

Walter, Chip. 2006. *Thumbs, Toes, and Tears and Other Traits that Make Us Human.* New York: Walker and Company.

Part V

Appendix

Appendix A

Are Some Organisms Poorly Designed?

The focus of this book has been on the design of the human body. What about the myriad of other organisms that we share the planet with?

A superficial study of the natural world helps one to become aware of many examples of animals and plants that *appear* to be both poorly designed and poorly adapted to their environments. Furthermore, many animals are able to survive in only a very narrow set of environmental conditions, and require a rigid specified ecological niche. Even small changes in environmental conditions can be lethal to many animals, and can even result in the extinction of an animal type. Many organisms are extremely fastidious in their nutritional requirements, and if these organisms were designed for their niches, the design seems poor. A focus here is, "Does evolution offer a better explanation than creationism for these putative 'poor design' features in nature?"

Consider the microbial world. Some bacteria can survive very adequately on only a few nutrient types, and others, as the *spiroplasmas*, are so fastidious that they require some 80 ingredients to survive.[1] Humans need only 10 different amino acids in their diet, and can make the other ten required. Some bacteria require a diet containing all 20 amino acids,

---

[1]Black, 1999.

yet other bacteria require only a few different amino acids. No pattern of primitive, less evolved to more evolved, is found.

One large, reddish-colored bacteria earned *The Guinness Book of World Records* "world's toughest bacterium" honor.[2] Named *Deinococcus radiodurans*, it was discovered in 1956 in a can of spoiled ground meat at the Corvallis, Oregon, Agricultural Experiment Station. The bacterium had even withstood the radiation used to sterilize the food, and has since been found to "tolerate one thousand times the radiation level that a person can." It can even live in the intense radiation of a nuclear reactor!

The *radiodurans* has "the remarkable ability to *realign* its radiation-shattered pieces of genetic material" and employs "enzymes to bring in new nucleotides from the outside environment and stitch together the pieces to repair the damage."[3] A question that must be asked is, "If this organism has evolved this critically important ability—which gives it a major survival advantage—why is this mechanism not more common?" This mechanism could virtually eliminate cancer and other genetic diseases among humans and animals. Why would it have been lost in the alleged macroevolutionary process, assuming this bacteria has an ancient heritage as do other bacteria?

At the macroscopic level, a good example of an extremely fastidious animal is the koala, which must subsist on a diet of only eucalyptus leaves. When eucalyptus leaves are in short supply, many koalas will starve even if many other types of edible food are in abundant supply. One of the best-known examples of an animal that was "poorly designed" for survival was the dodo bird. Its life-style and anatomy made it almost certain that it would become extinct if it encountered any aggressive, large, predatory animal.

The dodo (and all other non-flying birds) lays its eggs on flat ground instead of in a safer location. Laying eggs on the open ground exposes them to hundreds of ground-dwelling animals and, as a result, the eggs often are consumed. When rodents and other predators were introduced by sailors, the dodo soon became extinct. The ground egg-laying trait is an important reason why some birds are today threatened with extinction.

---

[2]Lewis, 2001, p. 157.
[3]Lewis, p. 157.

If they produced a large number of eggs, survival would be less of a problem, but many ground egg-laying birds lay only a few eggs, or even one egg, at a time.

Yet another well-known example is the giant panda. These animals are so inept at reproducing that only about a thousand pandas are left in the wild, in spite of a 30-year, multi-million-dollar campaign to encourage their breeding. Success has only recently been realized with new laboratory techniques. The reasons why they are threatened with extinction include the fact that they must subsist on only a single species of bamboo.[4] Their reproduction methods are also so inept that even under ideal conditions they do not reproduce very successfully, and under most conditions they don't reproduce at all.[5]

One would expect that millions of years of evolution would have honed their reproductive system to the point that they could effectively reproduce in their natural environment. Any minor improvement, no matter how small, would be selected for, and only a few changes would have made them much more fit. The same could be said of the koala and the dodo.

The examples of plants and animals that appear to be both poorly designed and poorly adapted to their environments can be explained by the reality that compromise in the natural world must exist in order for life to exist. Darwin's postulated macroevolution by natural selection will eventually cause the extinction of all life. Therefore, life must have built-in limits to ensure that balance is maintained and that one animal does not become too successful numerically. The example of cancer evolution is used to speculate on the eventual result of presumed natural selection; namely, the death of all life. This view of poor design is the dominant view in science today.

---

[4]Hsü, 1986, p. 49.
[5]Schaller, 1968.

Other Examples of Where Natural Selection Should Have Worked—But Hasn't

Crocodiles normally catch their prey by going to the water's edge, and then grabbing and drowning their victim. Certain types of deer-like mammals regularly drink by the water's edge, ignoring the local crocodiles that usually pull a deer into the water, to kill it by drowning before consuming it. After the millions of years claimed by evolutionists, it would seem that animals coming to water holes to drink where crocodiles feed would be able to effectively sense their presence better.

Those animals that are even *slightly* more aware of the crocodile's presence would be more likely to live and pass this trait on to their offspring. Eventually, the whole population would likewise be more effective in avoiding crocodiles. Neo-Darwinism also would predict that as a deer evolved to be more sensitive to crocodile noise, sight, and smell, the crocodile would evolve to be more discrete. Yet this has not happened. The deer are remarkably oblivious to the crocodiles, and the crocodiles need only to swim to where the deer are and then attack. As long as there are deer, the crocodiles will have plenty of food.

A recent example of what appears to be poor adaptation that would be strongly selected against is the male bean weevil's copulatory organ. It is a spine-covered structure that lacerates the females' copulatory organ. For obvious reasons, females typically fight potential mates by kicking with their hind legs.[6] As a result of the damage, females that never mate live much longer–about a month, whereas those that mate once live an average of ten days, and the twice-mated females live a mere nine days.

Any small mutation or genetic variant that reduces the stiffness or size of the spines would hypothetically be selected for, and it would seem that millions of years of evolution would have eliminated this major impediment to reproduction. Sooner or later, a mutation would occur that caused the male spines to be less rigid, or would cause their loss altogether. This mutant weevil would have increased its chances of mating significantly, and consequently this male would be more likely to have more offspring

---

[6]Crudgington and Siva-Jothy, 2000, p. 855.

than a bean weevil with the wild-type, rigid, spine-covered copulatory organ.

Another example is the need for dietary vitamin C, a critically important compound required for many body functions, not the least of which is as an antioxidant to help neutralize free radicals. Only a few animals cannot manufacture vitamin C (primarily carnivorous animals that do not feed on plant food, a major source of vitamin C). Because it often is difficult to obtain enough in the diet, the ability to synthesize vitamin C would confer to the animal a major survival advantage. Many so-called primitive organisms have this ability, but many higher animals lack it.

Evolutionists claim that this ability was lost during evolution into higher life forms. As purported evidence they cite a pseudogene (an inactive or damaged gene) involved in vitamin C production found in one sample (so far none has been found in any other primates). Yet, if this is the case, the activation of the gene would be highly favored. The animals that had the ability to manufacture vitamin C, would be able to survive in a far wider set of circumstances. Lack of vitamin C is understood today as a major cause of both a wide variety of diseases and abnormal conditions. No longer would a diet high in vitamin C such as citrus fruit be required, but an animal could do very well on a much poorer-quality diet.

Another example is the human species, *Homo sapiens*, supposedly the most highly evolved animal on Earth. Considering their body weight, humans are about ten times more vulnerable to toxins than are many experimental animals.[7] The difference is due partly to more effective biotransformation systems in many so-called lower animals and is an important reason why humans need to rigidly control their environment to survive.

Many animals also possess behavior traits that are often lethal—well-known examples are dogs and many other animals who commonly consume animal excrement. The reason why this behavior frequently is lethal is because around 40% of the dry weight of most mammal excretory matter is bacteria, many types of which are pathogenic.[8] As dumping untreated

---

[7]Klaassen, 1996, p. 27.
[8]Klaassen, 1996.

sewer water in drinking water can be disastrous, so too are the ways of many animals. Some Darwinists speculate that coprophagy (eating excrement) can be advantageous as a means of erasing the markings of a competitor for territorial reasons.

In fact, coprophagy rarely erases the scent, and most dogs do not limit their coprophagy to any one territory. Some studies indicate that coprophagy results from chronic stress[9] and otherwise is normally uncommon.[10] One study found it exists in about 9% of dogs, many of which were in stressful situations.[11] In the wild (where dogs face much more stress), coprophagy is evidently very common, but regardless of how common it is, coprophagy is still very harmful to health, and is not a functional response to stress. Surely, natural selection would have eliminated this trait after millions of years, or would never have selected for it.

Many examples of poor fitness, either of biochemistry or behavior, have been exploited by evolutionists as evidence that life was not created. They reason: "Why would a Creator create animals that were so obviously marginally or poorly adapted and can survive only in a very narrow ecological niche or environment?" Darwin claimed that these examples were evidences of poor design that militated against an Intelligent Design worldview. Conversely, since evolution is alleged to be an un-designed, undirected, and unplanned process, Darwinists reason that if evolution were true, it would not be unexpected to find many examples of poor design in nature.[12] The problem with this reasoning is, to

> find fault with biological design because it misses an idealized optimum, as Stephen Jay Gould regularly does, is therefore gratuitous. Not knowing the objectives of the designer, Gould is in no position to say whether the designer has come up with a faulty compromise among those objectives.[13]

The examples of poor design also argue against the efficacy of a mu-

---

[9]Beerda, et al., 1999.
[10]Crowell-Davis, et al., 1994.
[11]Wells and Hepper, 2000.
[12]Dawkins, 1996.
[13]Dembski, 2000, p. 2.

Figure A.1: Statue of Charles Darwin

Image Credit: 1000 Words / Shutterstock.com

tation/natural selection paradigm. Conversely, design theorists also have lacked a good explanation for these observations, except to point out that **Intelligent** Design is not necessarily **optimal** design, and that flaws in the creation are expected as a result of required compromises as well as the Biblical Fall.[14]

Evolutionists conclude that the reason poor design exists is because what evolves is a result of chance, time, and the constraints of natural law. If an adaptation works only well enough to allow an animal to survive and maintain its population size, the animal will not go extinct. But, in fact, evolution has major problems explaining what is commonly observed: millions of years of evolution should not have produced many poorly adapted, even inept, animals that can live for only a few hours or years. If an animal cannot successfully compete or survive, its traits will not be passed on to its offspring. Natural selection should, therefore, consistently select for the variations that can compete and function better. In the words of Timms and Read,

> factors that constrain niche expansion lie at the heart of a key problem in evolutionary ecology: why are there so many different types of species? Why is there not an ultimate organism adapted to exploit all ecological niches? ...Why are there no parasite species exploiting all the members of large taxa such as mammals or birds?[15]

Evolutionists attempt to answer this question by, for example, noting factors limiting a species' range, such as water barriers, for example. This may account for some cases, but another factor may be more important. As Timms and Read note: "We have remarkably little understanding of the relative importance of these alternatives in limiting host range in natural parasite populations" (p. 333).

Another example are water bears, crustaceans less than a millimeter long which are part of Phylum Tardigrada. The over four hundred species that have been identified inhabit a diversity of niches ranging from high

---

[14]Dembski, 2000.
[15]Timms and Read, 1999, p. 334.

mountains to the ocean abyss, and from the Arctic to the Antarctic. They can survive in temperatures ranging from higher than boiling water, to those as low as 0.0008 Kelvin, or close to absolute zero. These crustaceans survive environmental extremes by going into a profound dormancy state in which they are oblivious to hunger for hundreds of years, then awakening like Sleeping Beauty.

They also can withstand radiation a thousand times above the lethal dose for humans. They are, in many ways, extremely hardy, yet are inept in other ways, such as the ability to defend themselves against predators. Professor Hsü concluded that if in fact the "ability to survive a crisis is the bedrock criterion of fitness, then little water bears are the fittest of us all, and that is the direction, the purpose, and the perfection to which natural selection should have tended. Luckily, it has not."[16]

Actually, the best example of a "super-animal" may be human beings. We now have the ability to cause many, if not most, animals to become extinct. This power so far has not been exercised, partially because humans know that their life depends on the existence of a balanced ecosystem. Humans also normally have an innate instinctual love of animals, especially baby animals–although no doubt much of this is learned through culture.[17] Thus, knowledge, culture, and this putative internal instinct serve as a brake to enable humans to control their drive so as not to reach the state whereby all, or most, animals become extinct. Most recent extinctions were caused by humans or by natural disasters (such as the post-flood ice age) and not by other animals as a result of natural selection competition.

Why Survival of the Fittest Must Be Limited for the Survival of Life

Although most animals have their ecological niche, most animals nevertheless face some competition. This competition, though, must be controlled so that the proverbial "balance in nature" is maintained because,

---

[16]Hsü, 1986 p. 245.
[17]Marchant, 1968.

if it is lost, it must soon be reestablished or extinction results. To survive, therefore, natural selection cannot function to significantly improve the competition among various forms of life because by so doing will eventually cause the extinction of the competition, and eventually of all life.

Thus, just as many human inventions have built-in weak points that snap under pressure to prevent other points from failing, likewise built-in weakness must exist in all life in order to ensure that the balance of nature continues to exist. This built-in weakness can be interpreted as *necessary* in order to maintain a balance in nature, i.e., natural selection at best prunes out the inferior and weaker individuals, reducing the amount of devolution.

In industry, many machines contain a built-in designed weak point that will fail first, preventing more damage from occurring to other parts of the unit. The best examples are fuses and circuit breakers, which were designed to fail before the systems internal wires overheat enough to cause a fire or damage the electrical components. Fuses and circuit breakers have prevented millions of fires, and reduced, or prevented, the damage of multiplied millions of electrical and electronic equipment units.

Circuit breakers, considered one of the most important inventions ever, demonstrate intelligent design. Likewise, the "circuit breakers" found in nature that prevent one life kind from causing the extinction of other life forms also demonstrate intelligent design. This further illustrates the observation that **intelligent** design need not be **optimal** design.[18] This important, built-in innovation in life explains the major contradiction between the reality of nature and natural selection, and the balance found to exist in almost all areas as discovered by the study of ecology.

---

[18]Dembski, 2000.

Conclusions

The set of observations reviewed here have important implications for both the creation and evolution worldviews. They explain the observation that many animals–even the most intelligent ones–commonly manifest behavior that is inept from a survival standpoint. This view explains why designs once judged as imperfections in the natural world (obviously a misnomer, as is calling a fuse or circuit breaker an imperfection) actually have a critical function. This so-called imperfection is a necessary design required in order for life to survive in abundance and variety in the long-term. In spite of this built-in balance, occasionally the balance is offset, often due to human intervention, and occasionally due to major natural disasters, forcing a new equilibrium to be reached.

Intelligent Design theorists accept the fact that beneficial mutations prompting minimal information loss are possible (although extremely rare), and that natural selection has been documented to result in a limited level of improved adaptation to the local environment. The problem that Darwinists must address is the *origin* of the variation, not the fact that certain variations cannot facilitate individual survival.

Consequently, some evolution, called *microevolution*, is actually only *variation within the created kinds*, and not a problem for either I.D. theorists or creationists. Large-scale biologic change, or macroevolution, has never been shown to occur. Even if it could, such a change would be a grave threat to the ecological balance of the biosphere.

The view argued here is that, in addition to genetic mechanisms, eco-logical mechanisms also exist to prevent macroevolution. This is because, as is the case with cancer, elimination of the less well adapted life eventu-ally would result in extinction of all life. These mechanisms, both genetic and ecological, include features of nature that have been dismissed by evolutionists as "poor design." Christian creationists maintain that some of these "poor design" features may be a result of God's curse upon the world, and the introduction of death due to the fall.

Summary

The imperfections found in nature produce a balance that is *opposed* to the logical outcome of Darwin's concept of natural selection noted above. Natural selection, if carried to its logical outcome, would continue until sooner or later one "**super-species**" would take over the world. It then would be forced to fiercely compete with its own kind for food. This super-species not only would kill its direct competitors, but, in time, when edible plants became extinct from lack of nutrients previously supplied by animal life (especially the bacteria) they would be forced to devour each other until only one "super-animal" was left.

When the food chain was gone, this lone survivor then would die for want of food, forever ending all life on Earth. If animals constantly developed more effective reproduction and survival techniques, no matter how slowly, *eventually* this balance would be lost. In fact, we do not see this occurring, even in animals with short life spans and a large number of offspring. If natural selection were a major force, balance in any sphere of activity would actually be both a precarious and temporary situation, maintained for only a very short period of time.

References

Beerda, Bonne; Matthijs B.H. Schilder, Jan A.R.A.M. Van Hoof, Hans W. De Vries, and Jan A. Mol. "Chronic Stress in Dogs Subjected to Social and Spatial Restriction. I. Behavioral Responses." *Physiology & Behavior*, 66(2):233-242, April 1999.

Black, Jacquelyn. *Microbiology Principles and Explorations*. Saddle River, NJ: Prentice Hall, 1999.

Committee on Diet, Nutrition, and Cancer/Assembly of Life Sciences/National Research Council. *Diet, Nutrition, and Cancer*. Washington, D.C.: National Academy Press, 1982.

Crowell-Davis, S.L.; K. Barry, J.M. Ballam and D.P. Laflamme. 1995. "The Effect of Caloric Restriction on the Behavior of Pen-Housed Dogs: Transition from Restriction to Maintenance Diets and Long-Term Effects." *Applied Animal Behaviour Science*, 43:43-61.

Crudgington, Helen and Mike Siva-Jothy. "Genital Damage, Kicking and Early Death." *Nature*, 407:855, Oct. 19, 2000.

Dawkins, Richard. *The Selfish Gene*. NY and Oxford: Oxford University Press, 1976.

Dembski, William. "Intelligent design is not optimal design." *Metaviews*, Feb 2, 2000.

Demick, David. "Cancer and the Curse." *Back to Genesis*, 145:a-c, Jan 2001.

Hsü, Kenneth. *The Great Dying; Cosmic Catastrophe, Dinosaurs and the Theory of Evolution*. NY: Harcourt, Brace, Jovanovich, 1986.

Klaassen, Curtis. *Casarett and Doull's Toxicology*. McGraw Hill, 5th edition, 1996.

Lewis, Ricki. *Human Genetics*. NY: McGraw Hill, 2001.

Marchant, R. A. *Man and Beast*. NY: Macmillan, 1968.

Pauling, Linus. *Vitamin C the Common Cold and the Flu*. San Francisco, CA: W. H. Freeman and Company, 1976.

Schaller, George B. *The Last Panda*. Chicago, IL: University of Chicago Press, 1993.

Sullivan, G.F.; P.S. Amenta, J.D. Villanueva, C.J. Alvarez, J.M. Yang, and W.N. Hait. "The expression of drug resistance gene products during the progression of human prostate cancer." *Clinical Cancer Research*, 4(6):1393-403, 1998.

Timms, Rebecca and Andrew F. Read. "What makes a Specialist Special?" *Trends in Ecology & Evolution*, 14(9):333-334, September, 1999.

U.S. Department of Health and Human Services. The Surgeon General's Report on Nutrition and Health. U.S. Department of Health and Human Services: Government Printing Office, 1988.

Wentzler, Rich. *The Vitamin Book*. New York: Gramercy Publishing Company, 1978.

Weinberg, Robert. 1998. *One Renegade Cell; How Cancer Begins*. NY: Basic Books, 1998.

Wells, Deborah L. and Peter G. Hepper. "Prevalence of Behaviour Problems Reported by Owners of Dogs Purchased from an Animal Rescue Shelter." *Applied Animal Behaviour Science*, 69:55-65, 2000.

Appendix B

The Extent of the Acceptance of the Poor Design Argument

The extent of the acceptance of the poor design argument can be shown by the endorsements of Abby Hafer's book, *The Not-So-Intelligent Designer*, referenced in several chapters in the present volume, on the theme that humans are poorly designed. Note that these endorsements come from professors from some of the world's leading universities, including Harvard. This list embarrassingly illustrates the extent of how indoctrination can distort the mind.[1]

> I've been dreaming of a politically edgy treatment of intelligent design and here it is at last. Abby Hafer is acutely intelligent and wonderfully witty. Read this book and laugh your way to clarity and wisdom.
>
> —**Wesley J. Wildman**
> Boston University, Boston, MA

---

[1] https://wipfandstock.com/the-not-so-intelligent-designer.html.

Three cheers for Abby Hafer! She did it and no one thought it could be done! She wrote a devastating critique of intelligent design that is clear, funny, scientifically accurate, and charming. Her book is a marvel of how popular science should be written. Oh, were there more scientific writers like Abby . . .

—Michael Martin
Boston University, Boston, MA

A delightful exploration of the quirks of our bodies that make biology so much fun, evolution so fascinating, life so explicable, and intelligent design creationism so preposterous.

—Steven Pinker
Harvard University, Cambridge, MA

Where has this book been all my life? This work by Dr. Hafer systematically overturns the arguments of the intelligent design movement with wit and plain language. As a pastor, I appreciate Hafer's contribution to clarity in our public discourse, both scientific and political. Her intention may be to restore science to its rightful place, but she has also done the faith community a favor, liberating it from a silly and unnecessary controversy.

—Julia Tipton Rendon
Crossroads United Church of Christ, Indianola, IA

For an adequate account of the world, we must take a sober look at life as it really is. Hafer shows that things are a whole lot messier and makeshift than what some intelligent design theories would incline us to believe. This book has the potential not only to alter the political terrain in wars over evolution and creationism but also to prompt believers like me to rethink how we should talk about God as Creator.

—**Thomas Jay Oord**
Author of *Divine Grace and Emerging Creation*

The Not-So-Intelligent Designer is a much-needed work in an America where anti-intellectualism is rampant and, shockingly, even candidates for high office frequently reject evolution. Abby Hafer has that rare ability to communicate complex scientific ideas in understandable terms for non-scientists, and this book is sure to enlighten many.

—**David Niose**
Author of *Fighting Back the Right*

Hafer's book is a valuable contribution to debunking the claims of intelligent design and the notion of one or more gods intervening in the physics and biology of the real world. She writes in an engaging style that entertains as well as informs. I enthusiastically recommend it.

—**Ellery Schempp**
Brown University, Providence, RI

Intelligent design creationism is a dangerously successful political ideology—they've passed laws, co-opted high school teachers, and nearly half the population identifies as creationist. *The Not So Intelligent Designer* is a guidebook on how intelligent design fails, from unintelligently designed testes to intelligent design's unconstitutional religious agenda.

—Zack Kopplin
Organizer of the campaign to repeal the Louisiana
Science Education Act, Baton Rouge, LA

The Not-So-Intelligent Designer is a scholarly book that is accessible and intelligible to the general reader. It is especially a must read for adherents of religious traditions who embrace modern science. Hafer does a masterful job of defining science, i.e. a way of knowing characterized by the formulation of a hypothesis, the gathering of evidence, drawing of conclusions, and repeating the experiment. She shows in a compelling way that the conclusions of advocates of intelligent design are beyond the purview of science. In fact, she provides a convincing demonstration of how the battles over intelligent design are over the nature of science itself.

—Leslie A. Muray
Curry College, Milton, MA